LABORATORY MANUAL FOR

CHM 1010

Wright State University
Department of Chemistry

EDITORS

Michelle Newsome
Kirby A. Underwood

Assembled from contributions of the Faculty of the Department of Chemistry.

Introduction to Chemistry

Course instructor/contact Information: _______________________

Course TA(s)/contact information: _______________________

Fire extinguisher location: _______________________

Shower and eye wash location: _______________________

Nearest telephone location: _______________________

Telephone number for aid: _______________________

Fire escape location from laboratory: _______________________

Nearest AED location: _______________________

VAN-GRINER
LEARNING

Laboratory Manual for

CHM 1010: Introduction to Chemistry

Fourth Edition
Editors: Travis B. Clark, Michelle L. Newsome, and Kirby A. Underwood
Wright State University
Department of Chemistry

Printed in the United States of America
10 9 8 7 6 5 4 3 2
ISBN: 978-1-61740-808-3

Van-Griner Learning
Cincinnati, Ohio
www.van-griner.com

President: Dreis Van Landuyt
Project Manager: Brenda Schwieterman
Customer Care Lead: Lauren Houseworth

Newsome 808-3 Su19
313567-319931
Copyright © 2020

Acknowledgments

The editors are indebted to the writers of all previously published laboratory manuals, the specific contributions of whom are no longer traceable.

Special thanks are due to the members of the Chemistry Department of Wright State University for their comments and suggestions during the preparation of this manual.

We would also like to thank numerous laboratory assistants, who provided their own comments and suggestions.

Table of Contents

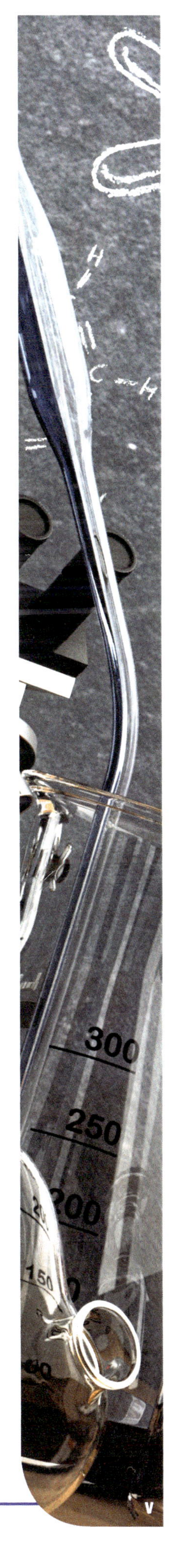

Safety Guidelines
1010 Laboratories

WHAT ARE THE SAFETY DOs AND DON'Ts FOR STUDENTS?

Life threatening injuries can happen in the laboratory. For that reason, students need to be informed of the correct way to act and things to do in the laboratory. The following is a safety checklist to acquaint students with the safety dos and don'ts in the laboratory.

Conduct

- Do not engage in practical jokes or boisterous conduct in the laboratory.
- Never run in the laboratory.
- The use of personal audio or video equipment is prohibited in the laboratory.
- The performance of unauthorized experiments is strictly forbidden.
- Do not sit on laboratory benches.

General Work Procedure

- Know emergency procedures.
- Never work in the laboratory without the supervision of a TA.
- Always perform the experiments or work precisely as directed by the TA.
- Immediately report any spills, accidents, or injuries to a TA.
- Never leave experiments while in progress.
- Never attempt to catch a falling object.

- Be careful when handling hot glassware and apparatus in the laboratory. Hot glassware looks just like cold glassware.
- Never point the open end of a test tube containing a substance at yourself or others.
- Never fill a pipette using mouth suction. Always use a pipetting device.
- Make sure no flammable solvents are in the surrounding area when lighting a flame.
- Do not leave lit Bunsen burners unattended.
- Turn off all heating apparatus, gas valves, and water faucets when not in use.
- Do not remove any equipment or chemicals from the laboratory.
- Coats, bags, and other personal items must be stored in designated areas, not on the bench tops or in the aisle ways.
- Notify your instructor or TA of any sensitivities that you may have to particular chemicals if known.
- Keep the floor clear of all objects (e.g., ice, small objects, spilled liquids).

Housekeeping

- Keep work area neat and free of any unnecessary objects.
- Thoroughly clean your laboratory work space at the end of the laboratory session.
- Do not block the sink drains with debris.
- Never block access to exits or emergency equipment.
- Inspect all equipment for damage (cracks, defects, etc.) prior to use; do not use damaged equipment.
- Never pour chemical waste into the sink drains or wastebaskets.
- Place chemical waste in appropriately labeled waste containers.
- Properly dispose of broken glassware and other sharp objects (e.g., syringe needles) immediately in designated containers.
- Properly dispose of weigh boats, gloves, filter paper, and paper towels in the laboratory.

Apparel in the Laboratory

- Always wear appropriate eye protection (i.e., chemical splash goggles) in the laboratory.
- Wear disposable gloves, as provided in the laboratory, when handling hazardous materials. Remove the gloves before exiting the laboratory.
- Wear a full-length, long-sleeved laboratory coat or chemical-resistant apron.
- Wear shoes that adequately cover the whole foot; low-heeled shoes with non-slip soles are preferable. Do not wear sandals, open-toed shoes, open-backed shoes, or high-heeled shoes in the laboratory.

- Avoid wearing shirts exposing the torso, shorts, or short skirts; long pants that completely cover the legs are preferable.
- Secure long hair and loose clothing (especially loose long sleeves, neck ties, or scarves).
- Remove jewelry (especially dangling jewelry).
- Synthetic finger nails are not recommended in the laboratory; they are made of extremely flammable polymers which can burn to completion and are not easily extinguished.

Hygiene Practices

- Keep your hands away from your face, eyes, mouth, and body while using chemicals.
- Food and drink, open or closed, should never be brought into the laboratory or chemical storage area.
- Never use laboratory glassware for eating or drinking purposes.
- Do not apply cosmetics while in the laboratory or storage area.
- Wash hands after removing gloves and before leaving the laboratory.
- Remove any protective equipment (i.e., gloves, lab coat or apron, chemical splash goggles) before leaving the laboratory.

Emergency Procedure

- Know the location of all the exits in the laboratory and building.
- Know the location of the emergency phone.
- Know the location of and know how to operate the following:
    - Fire extinguishers
    - Alarm systems with pull stations
    - Fire blankets
    - Eye washes
    - First-aid kits
    - Deluge safety showers
- In case of an emergency or accident, follow the established emergency plan as explained by the TA and evacuate the building via the nearest exit.

Chemical Handling

- Check the label to verify it is the correct substance before using it.

- Wear appropriate chemical resistant gloves before handling chemicals. Gloves are not universally protective against all chemicals.

- If you transfer chemicals from their original containers, label chemical containers as to the contents, concentration, hazard, date, and your initials.

- Always use a spatula or scoopula to remove a solid reagent from a container.

- Do not directly touch any chemical with your hands.

- Never use a metal spatula when working with peroxides. Metals will decompose explosively with peroxides.

- Hold containers away from the body when transferring a chemical or solution from one container to another.

- Use a hot water bath to heat flammable liquids. Never heat directly with a flame.

- Add concentrated acid to water slowly. Never add water to a concentrated acid.

- Weigh out or remove only the amount of chemical you will need. Do not return the excess to its original container, but properly dispose of it in the appropriate waste container.

- Never touch, taste, or smell any reagents.

- Never place the container directly under your nose and inhale the vapors.

- Never mix or use chemicals not called for in the laboratory exercise.

- Use the laboratory chemical hood, if available, when there is a possibility of release of toxic chemical vapors, dust, or gases. When using a hood, the sash opening should be kept at a minimum to protect the user and to ensure efficient operation of the hood. Keep your head and body outside of the hood face. Chemicals and equipment should be placed at least six inches within the hood to ensure proper air flow.

- Clean up all spills properly and promptly as instructed by the teacher.

- Dispose of chemicals as instructed by the TA.

- When transporting chemicals (especially 250 mL or more), place the immediate container in a secondary container or bucket (rubber, metal, or plastic) designed to be carried and large enough to hold the entire contents of the chemical.

- Never handle bottles that are wet or too heavy for you.

- Use equipment (glassware, Bunsen burner, etc.) in the correct way, as indicated by the TA.

WHAT IS A MATERIAL SAFETY DATA SHEET?

Material Safety Data Sheets (MSDS) contain information regarding the proper procedures for handling, storing, and disposing of chemical substances.

- An MSDS accompanies all chemicals or kits that contain chemicals.

- If an MSDS does not accompany a chemical, many web sites and science supply companies can supply one or they can be obtained from *www.msdsonline.com.*

- Typically the information is listed in a standardized format (ANSI Z400.1-1998, Hazardous Industrial Chemicals-Material Safety Data Sheet-Preparation).

- Refer to sections *General Guidelines to Follow in the Event of a Chemical Accident or Spill* and *Understanding an MSDS* in this front matter for additional information on the format and content of MSDSs (ANSI format).

COMMON SAFETY SYMBOLS

The United Nation's Globally Harmonized System (GHS) of classification and labeling of chemicals.

<table>
<tr><td>

Health Hazard

- Carcinogen
- Mutagenicity
- Reproductive toxicity
- Respiratory sensitizer
- Target organ toxicity
- Aspiration toxicity

</td><td>

Flame

- Flammables
- Pyrophorics
- Self-heating
- Emits flammable gas
- Self-reactives
- Organic peroxides

</td><td>

Exclamation Mark

- Irritant (skin and eye)
- Skin sensitizer
- Acute toxicity (harmful)
- Narcotic effects
- Respiratory tract irritant
- Hazardous to ozone layer (non-mandatory)

</td></tr>
<tr><td>

Gas Cylinder

- Gases under pressure

</td><td>

Corrosive

- Skin corrosion/burns
- Eye damage
- Corrosive to metals

</td><td>

Exploding Bomb

- Explosives
- Self-reactives
- Organic peroxides

</td></tr>
<tr><td>

Flame Over Circle

- Oxidizers

</td><td>

Environment
(non-mandatory)

- Aquatic toxicity

</td><td>

Skull and Crossbones

- Acute toxicity (fatal or toxic)

</td></tr>
</table>

NATIONAL FIRE PROTECTION ASSOCIATION HAZARD LABELS

The National Fire Protection Association (NFPA) has developed a visual guide (right) for a number of chemicals pertinent to the MSDS. The ANSI/ NFPA 704 Hazard Identification system, the NFPA diamond, is a quick visual review of the health hazard, flammability, reactivity, and special hazards a chemical may present.

The diamond is broken into four sections (blue, red, yellow, and white). The symbols and numbers in the four sections indicate the degree of hazard associated with a particular chemical or material.

	Health Hazard (blue)	
4	Danger	May be fatal on short exposure. Specialized protective equipment required.
3	Warning	Corrosive or toxic. Avoid skin contact or inhalation.
2	Warning	May be harmful if inhaled or absorbed
1	Caution	May be irritating
0		No unusual hazard

	Flammability (red)	
4	Danger	Flammable gas or extremely flammable liquid
3	Warning	Combustible liquid flash point below 100°F
2	Caution	Combustible liquid flash point of 100° to 200°F
1		Combustible if heated
0		Not combustible

	Reactivity (yellow)	
4	Danger	Explosive material at room temperature
3	Danger	May be explosive if shocked, heated under confinement or mixed with water
2	Warning	Unstable or may react violently if mixed with water
1	Caution	May react if heated or mixed with water but not violently
0	Stable	Not reactive when mixed with water

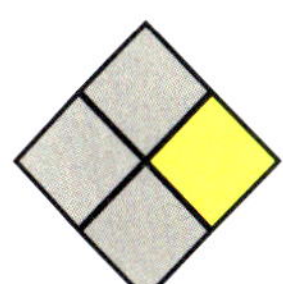

	Special Notice Key (white)	
W	Water Reactive	
OX	Oxidizing Agent	

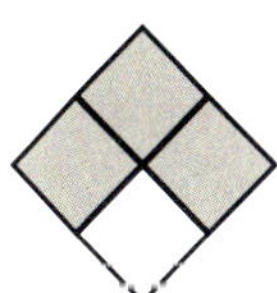

HOW DOES A CHEMICAL ENTER THE BODY?

A chemical can enter the body through different routes.

- These different routes of exposure and the types of exposure (acute or chronic) can affect the toxicity of the chemical.
- The most probable (primary) route(s) of exposure to a chemical will be identified in the MSDS.
- Three principal routes of exposure include dermal exposure (skin), inhalation, and ingestion (oral).

Dermal Exposure

Although the skin is an effective barrier for many chemicals, it is a common route of exposure. The toxicity of a chemical depends on the degree of absorption that occurs once it penetrates the skin. Once the skin is penetrated, the chemical enters the blood stream and is carried to all parts of the body. Chemicals are absorbed much more readily through injured, chapped, or cracked skin, or needle sticks than through intact skin. Generally, organic chemicals are much more likely to penetrate the skin than inorganic chemicals.

Dermal exposure to various substances can also cause irritation and damage to the skin and/or eyes. Depending on the substance and length of exposure, effects of dermal exposures can range from mild temporary discomfort to permanent damage.

Inhalation

Inhalation is another route of chemical exposure. Chemicals in the form of gases, vapors, mists, fumes, and dusts entering through the nose or mouth can be absorbed through the mucous membranes of the nose, trachea, bronchi, and lungs. Unlike the skin, lung tissue is not a very protective barrier against the access of chemicals into the body. Chemicals, especially organic chemicals, enter into the blood stream quickly. Chemicals can also damage the lung surface.

Ingestion

Ingestion involves chemicals entering the body through the mouth. Chemical dusts, particles, and mists may be inhaled through the mouth and swallowed.

They may also enter through contaminated objects, such as hands or food that come in contact with the mouth. Absorption of the chemicals into the bloodstream can occur anywhere along the length of the gastrointestinal (GI) tract.

WHAT ARE EXPOSURE LIMITS?

Exposure limits are intended to protect workers from excessive exposure to hazardous substances:

- Established by health and safety authorities and chemical manufacturers
    - Department of Labor's Occupational Safety and Health Administration (OSHA)
    - American Conference of Governmental Industrial Hygienists (ACGIH)
    - National Institute for Occupational Safety and Health (NIOSH)
    - Environmental Protection Agency (EPA)
    - American Industrial Hygiene Association (AIHA)
- Define the amount/concentration to which a worker can be exposed without causing an adverse health effect.
- Typically pertain to the concentration of a chemical in the air, but may also define limits for physical agents such as noise, radiation, and heat.
- Usually can be found on the MSDS; make sure your MSDSs are up-to-date.

Exposure Limits

Legally Enforceable Limits

Permissible Exposure Limits (PELs)

- Set by OSHA, 29 CFR 1910.1000, and 1910.1001 through 1910.1450.
- Specifies the maximum amount or concentration of a chemical to which a worker may be exposed.
- Generally defined in three different ways:
    - **Ceiling Limit (C):** The concentration that must not be exceeded at any part of the workday
    - **Short Term Exposure Limit (STEL):** The maximum concentration to which workers may be exposed for a short period of time (15 minutes)
    - **Time Weighted Average (TWA):** The average concentration to which workers may be exposed for a normal, 8-hour workday

Other U.S. Exposure Limits

Threshold Limit Values (TLVs)

- Prepared by ACGIH volunteer scientists
- Denotes the level of exposure that nearly all workers can experience without an unreasonable risk of disease or injury
- An advisory limit; not enforceable by law
- Generally can be defined as ceiling limits, short term exposure limits, and/or time-weighted averages
- Usually equivalent to PELs

Recommended Exposure Limits (RELs)

- Recommended by NIOSH
- Indicates the concentration of a substance to which a worker can be exposed for up to a 10-hour workday during a 40-hour work week without adverse effects, however, sometimes based on technical feasibility
- Based on animal and human studies
- Generally expressed as a ceiling limit, short-term exposure limit, or a time-weighted average
- Often more conservative than PELs and TLVs

Workplace Environmental Exposure Limits (WEELs)

- Developed by AIHA volunteers
- Advisory limits; not enforceable by law
- Typically developed for chemicals that are not widely used or for which little toxicity information is available

Company-Developed Limits

- Developed by company scientists
- Advisory limits; not enforceable by law
- Usually based on only short-term studies of animals
- Generally intended for internal company use and sometimes for the customers

GENERAL GUIDELINES TO FOLLOW IN THE EVENT OF A CHEMICAL ACCIDENT OR SPILL

- Assess the overall situation.
- Determine the appropriate action to resolve the situation.
- Follow the pre-existing, approved local emergency plan.
- Act swiftly and decisively.

Below are some recommended actions for specific emergencies. Some of the actions have been proposed by the Council of State Science Supervisors in Science and Safety: Making the Connection.

Chemical in the Eye

- Flush the eye immediately with water while holding the eye open with fingers.
- If wearing contact lens, remove and continue to rinse the eye with water.
- Continue to flush the eye and seek immediate medical attention.

Acid/Base Spill

For a spill not directly on human skin, do the following:

- Neutralize acids with powdered sodium hydrogen carbonate (sodium bicarbonate/baking soda), or bases with vinegar (5% acetic acid solution).
- Avoid inhaling vapors.
- Spread diatomaceous earth to absorb the neutralized chemical.
- Sweep up and dispose of as hazardous waste.

For spills directly on human skin, do the following:

- Flush area with copious amounts of cold water from the faucet or drench shower for at least 5 minutes.
- If spill is on clothing, first remove clothing from the skin and soak the area with water as soon as possible.
- Arrange treatment by medical personnel.

UNDERSTANDING AN MSDS

ANSI Standardized MSDS Format

Section 1 gives details on *what the chemical or substance is, CAS number, synonyms, the name of the company issuing the data sheet,* and often an *emergency contact number.*

Section 2 identifies the *OSHA hazardous ingredients,* and may include *other key ingredients* and exposure limits.

Section 3 lists the major *health effects* associated with the chemical. Sometimes both the acute and chronic hazards are given.

Section 4 provides *first aid measures* that should be initiated in case of exposure.

Section 5 presents the *fire-fighting measures* to be taken.

Section 6 details the *procedures to be taken in case of an accidental release.* The instructions given may not be sufficiently comprehensive in all cases, and local rules and procedures should be utilized to supplement the information given in the MSDS sheet.

Section 7 addresses the *storage and handling* information for the chemical. This is an important section as it contains information on the flammability, explosive risk, propensity to form peroxides, and chemical incompatibility for the substance. It also addresses any special storage requirements for the chemical (i.e., special cabinets or refrigerators).

Section 8 outlines the *regulatory limits for exposure,* usually the maximum permissible exposure limits (PEL) (refer to section *How Does a Chemical Enter the Body?* of the front matter). The PEL, issued by the Occupational Safety and Health Administration, tells the concentration of air contamination a person can be exposed to for 8 hours a day, 40 hours per week over a working lifetime (30 years) without suffering adverse health effects. It also provides information on personal protective equipment.

Section 9 gives the *physical and chemical properties* of the chemical. Information such as the evaporation rate, specific gravity, and flash points are given.

Section 10 gives the *stability and reactivity* of the chemical with information about chemical incompatibilities and conditions to avoid.

Section 11 provides both the *acute and chronic toxicity* of the chemical and any health effects that may be attributed to the chemical.

Section 12 identifies both the *ecotoxicity* and the environmental fate of the chemical.

Section 13 offers suggestions for the *disposal of the chemical.* Local, State, and Federal regulations should be followed.

Section 14 gives the *transportation information* required by the Department of Transportation. This often identifies the dangers associated with the chemical, such as flammability, toxicity, radioactivity, and reactivity.

Section 15 outlines the *regulatory information* for the chemical. The hazard codes for the chemical are given along with principle hazards associated with the chemical. A variety of country and/or state specific details may be given.

Section 16 provides *additional information* such as the label warnings, preparation and revision dates, name of the person or firm that prepared the MSDS, disclaimers, and references used to prepare the MSDS.

SAMPLE MSDS

Material Safety Data Sheet

Toluene   *MSDS No. XXXX*

1. Product and Company Identification

Product Name: Toluene

Synonyms: Methylbenzene, Methylbenzol, Phenylmethane, Toluol

CAS No.: 108–88–3

Chemical Formula: $C_6H_5-CH_3$

Catalog Number: Tol 12

Supplier: Company X

        XXXXXXXXXXX

        Anywhere, XX XXXXX

        Emergency Information: 800-XXX-XXXX

2. Composition/Information on Ingredients

Ingredient	CAS No	Percent	Hazardous
Toluene	108–88–3	100%	Yes

3. Hazards Identification

Emergency Overview

Danger! Harmful or fatal if swallowed. Vapor harmful. Poison! May be absorbed through intact skin. Flammable liquid and vapor. May cause liver and kidney damage, may affect blood system or central nervous system. Causes irritation to skin, eyes, and respiratory tract.

Potential Acute Health Effects

Eye Contact: Causes severe eye irritation with redness and pain.

Skin Contact: Causes irritation. May be absorbed through skin.

Inhalation: Inhalation may cause irritation of the upper respiratory tract. Symptoms of overexposure may include fatigue, confusion, headache, dizziness, and drowsiness. Very high concentrations may cause unconsciousness and death.

Ingestion: Swallowing may cause abdominal spasms and other symptoms that parallel over-exposure from inhalation. Aspiration of material into the lungs may cause chemical pneumonitis, which may be fatal.

Chronic Exposure: Chronic exposure may result in anemia, decreased blood cell count, and bone marrow hypoplasia. Liver and kidney damage may occur. Repeated or prolonged contact may cause dermatitis.

4. First Aid Measures

Eye Contact: Immediately flush eyes with plenty of water for at least 15 minutes, lifting the upper and lower eyelids occasionally. Get medical attention immediately.

Skin Contact: In case of contact, immediately flush skin with plenty of soap and water for at least 15 minutes while removing contaminated clothing and shoes. Wash clothing before reuse. Call a physician immediately.

Inhalation: Evacuate victim to fresh air immediately. If not breathing, give artificial respiration. If breathing is difficult, give oxygen. Seek medical aid immediately.

Ingestion: Aspiration hazard. If swallowed, ***do not induce vomiting.*** Give 2–4 cups of milk or water. Never give anything by mouth to an unconscious person. Get medical attention immediately.

5. Fire Fighting Measures

Fire: Flash point: 4°C (40°F)

Autoignition temperature: 480°C (896°F)

Flammable limits in air % by volume: lower: 1.3%; upper: 7.1%

Flammable liquid and vapor!

Extremely flammable when exposed to flame or sparks. Vapors are heavier than air and can flow along surfaces to distant ignition source and flash back.

Explosion: Vapor-air concentrations above flammable limits are explosive. Contact with strong oxidizers may cause fire or explosion. Sensitive to static discharge.

Fire Extinguishing Media: Dry chemical, carbon dioxide, or foam. Material is lighter than water and a fire may be spread by use of water. Water may be used to cool fire surface and protect personnel. Water may also be used to flush spills away from exposures and to dilute spills to non-flammable mixtures. Avoid flushing hydrocarbon into sewers.

Special Information: In the event of a fire, wear full protective clothing and NIOSII-approved self-contained breathing apparatus operated in the pressure demand or other positive pressure mode.

6. Accidental Release Measures

Avoid Contact: Ventilate area of leak or spill. Remove all ignition sources. Wear appropriate personal protective equipment as specified in Section 8 Isolate hazard area. Contain and recover liquid when possible. Collect liquid in an appropriate container or absorb with an inert material such as earth, sand, or vermiculite. Do not use combustible materials, such as saw dust. Do not flush to sewer.

7. Handling and Storage

Handling: Wash thoroughly after handling. Use with adequate ventilation. Avoid contact with skin, eyes, or clothes. Electrically ground and bond containers when transferring material to avoid static accumulation.

Storage: Store in a cool, dry, well-ventilated location, away from any area where the fire hazard. Separate from incompatibles. Storage and use areas should be No Smoking areas. Use non-sparking type tools and equipment, including explosion proof ventilation. Containers of this material may be hazardous when empty since they retain product residues (vapors, liquid). Observe all warnings and precautions listed for the product. Protect container against physical damage. Keep container tightly closed.

8. Exposure Controls/Personal Protection

Ventilation System: A system of local and/or general exhaust is recommended to keep exposures below the Airborne Exposure Limits.

Exposure Limits: Toluene:

- OSHA Permissible Exposure Limit (PEL): 200 ppm TWA; 300 ppm (acceptable ceiling conc.); 500 ppm (acceptable maximum conc.)
- NIOSH Recommended Exposure Limit (REL): 100 ppm TWA (375 mg/m^3); STEL 150 ppm (560 mg/m^3)
- ACGIH Threshold Limit Value (TLV): 50 ppm TWA skin—potential for cutaneous absorption

Personal Respirators (NIOSH/EN 149 Approved): If the exposure limit is exceeded a half-face organic vapor respirator may be worn for up to ten times the exposure limit. A full-face organic vapor respirator or self-contained breathing apparatus may be worn up to 50 times the exposure limit. For emergencies or instances where the exposure levels are not known, use a full-face piece positive-pressure, air-supplied respirator.

Skin Protection: Wear impervious protective clothing, including boots, gloves, lab coat, apron or coveralls, as appropriate, to prevent skin contact.

Eye Protection: Use chemical splash goggles and/or a full face shield. Maintain eyewash fountain facilities in work area.

9. Physical and Chemical Properties

Physical State and Appearance: Clear, colorless liquid

Odor: Aromatic benzene-like

Solubility: Very slight

Specific Gravity (Water = 1): 0.9

Viscosity: 20cP @ 20°C

Boiling Point: 110°C (232°F)

Melting Point: –95°C (–139°F)

Vapor Density (Air = 1): 3.1

Vapor Pressure (mmHg): 53.3 @ 20°C (68°F)

Evaporation Rate (Butyl acetate = 1): 2.4

Molecular Formula: $C_6H_5CH_3$

Molecular Weight: 92.06

10. Stability and Reactivity

Stability: Stable under ordinary conditions of use and storage. Containers may burst when heated.

Hazardous Decomposition Products: Carbon dioxide and carbon monoxide may form when heated to decomposition.

Hazardous Polymerization: Has not been reported.

Incompatibilities: Heat, flame, strong oxidizers, nitric and sulfuric acids; will attack some forms of plastics, rubber, coatings.

Conditions to Avoid: Heat, flames, ignition sources and incompatibles.

11. Toxicological Information

Toxicological Data

Oral rat LD_{50}: 636 mg/kg

Skin rabbit LD_{50}: 14100 uL/kg

Inhalation rat LC_{50}: 49 gm/m^3/4H

Inhalation mouse LC_{50}: 400 ppm/24H

Irritation data: skin rabbit, 500 mg, Moderate

Eye rabbit, 2 mg/24H, Severe

Investigated as a tumorigen, mutagen, reproductive effector.

Reproductive Toxicity

Has shown some evidence of reproductive effects in laboratory animals.

12. Ecological Information

Environmental Fate: When released into the soil, this material may evaporate and is micro biologically biodegradable. When released into the soil, this material is expected to leach into groundwater. When released into water, this material may evaporate and biodegrade to a moderate extent. When released into the air, this material may be moderately degraded by reaction with photochemically produced hydroxyl radicals.

Environmental Toxicity: No data available; however, this material is expected to be toxic to aquatic life.

13. Disposal Considerations

Waste material should be handled as hazardous waste and sent to a RCRA approved incinerator or disposed in a RCRA approved waste facility. Processing, use, or contamination of this product may change the waste management options. State and local disposal regulations may differ from Federal disposal regulations. Dispose of container and unused contents in accordance with Federal, State, and local requirements.

14. Transport Information

Domestic (Land, U.S. D.O.T.)

Proper Shipping Name: Toluene

Hazard Class: 3

UN/NA: UN1294

Packing Group: II

Canada TDG

Proper Shipping Name: Toluene

Hazard Class: 3 (9.2)

UN/NA: UN1294

Packing Group: II

Additional Information: Flashpoint 4 C

15. Regulatory Information

CALIFORNIA PROPOSITION 65: WARNING

This product contains a chemical known to the State of California to cause birth defects or other reproductive harm.

Reportable Quantity: 1000 Pounds (454 Kilograms) (138.50 Gals)

NFPA Rating: Health – 2; Fire – 3; Reactivity – 0 (0 = Insignificant; 1 = Slight; 2 = Moderate; 3 = High; 4 = Extreme)

Carcinogenicity Lists: No

NTP: No

IARC Monograph: No

OSHA Regulated: No

Section 313 Supplier Notification: This product contains the following toxic chemical(s) subject to the reporting requirements of SARA TITLE III Section 313 of the Emergency Planning and Community Right-To-Know Act of 1986 and of 40 CFR 372:

CAS No.	Chemical Name	% by Weight
108–88–3	Toluene	100

16. Other Information

Label Hazard Warning

POISON! DANGER! HARMFUL OR FATAL IF SWALLOWED. HARMFUL IF INHALED OR ABSORBED THROUGH SKIN. VAPOR HARMFUL. FLAMMABLE LIQUID AND VAPOR. MAY AFFECT LIVER, KIDNEYS, BLOOD SYSTEM, OR CENTRAL NERVOUS SYSTEM. CAUSES IRRITATION TO SKIN, EYES, AND RESPIRATORY TRACT.

Label Precautions

- Keep away from heat, sparks, and flame. Keep container closed.
- Use only with adequate ventilation. Wash thoroughly after handling.
- Avoid breathing vapor.
- Avoid contact with eyes, skin, and clothing.

Label First Aid

Aspiration hazard. If swallowed, ***do not induce vomiting.*** Give large quantities of water. Never give anything by mouth to an unconscious person. If vomiting occurs, keep head below hips to prevent aspiration into lungs. If inhaled, remove to fresh air. If not breathing, give artificial respiration. If breathing is difficult, give oxygen. In case of contact, immediately flush eyes or skin with plenty of water for at least 15 minutes. Remove contaminated clothing and shoes. Wash clothing before reuse. In all cases call a physician immediately.

References: Upon request

Messing with Mixtures

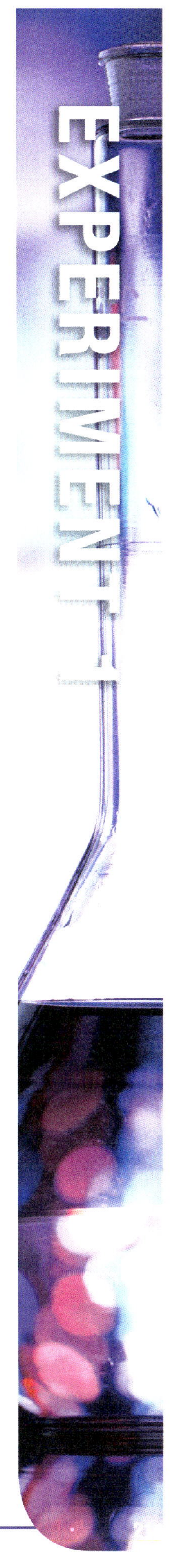

Objectives

- The students will be able to separate various types of mixtures through physical means.
- The students will identify mixtures based on physical properties.

Safety Precautions and Hazards

- This experiment requires the use of a variety of materials and substances. Students should be made aware of how to conduct themselves responsibly when touching and observing the items. The items should not be eaten in science class or spilled on the floor to avoid unnecessary hazards.

PRE-LAB QUESTIONS

Name: _______________________________________   Partners: _______________________________________

TA: _______________________________________   Section: _______________   Date: _______________

Answer the following questions using information obtained from lecture, the textbook, or lab manual. The responses will be checked at the start of your lab section for credit. Failure to complete may result in exclusion from participating in the lab.

1. The two major categories of chemicals are _______________________ and
   _______________________ .

2. _______________________ cannot be broken down into simpler substances by chemical means.

3. _______________________ can be broken down into simpler substances using chemical methods.

4. _______________________ contain two or more atoms connected to one another with chemical bonds.

5. Chemical properties of a substance are properties that describe what?

6. List 3–4 examples of physical properties of a substance.

7. In the following reaction,

   $$2H^+(aq) + CaCO_3(s) \rightarrow Ca^{2+}(aq) + H_2O(l) + CO_2(g)\uparrow$$

   the "↑" indicates that the product is _______________________ .

8. In the following reaction,

   $$Ca^{2+}(aq) + CO_3{}^{2-}(aq) \rightarrow CaCO_3(s)\downarrow$$

   the "↓" indicates that the product is a _______________________ .

PROCEDURE

Part A: Hit the Trail!

Record your observations on the Data Sheet.

 9. Look at the mixture in Bag A. What do you see?

10. Would this mixture be classified as a heterogeneous or homogeneous mixture? Give a reason for your answer.

11. What is the total mass of your mixture? Be sure to subtract the mass of the Ziploc bag. Record this amount in the "mass of mixture" column in Table 1.1.

12. Separate the parts of the mixture and find the mass of each group. Use the formula below to calculate the percentage for each part of the mixture. Record your data in Table 1.1.

Mass of Substance (g) ÷ Mass of Mixture (g) × 100

Part B: Tasty Solutions

Read the steps below, and then obtain three pieces of M&M candy from your teacher. You will need three pieces for each group member. Record your observations on the Data Sheet.

13. Place one piece of candy in your mouth and allow it to dissolve *without using your tongue or teeth to help!* In Table 1.2, record the time (in seconds) it takes for the **candy shell** to dissolve.

14. Place another piece of candy in your mouth and allow it to dissolve using *only your tongue* to move it around. Record the time (in seconds) it takes for the **candy shell** to dissolve.

15. Place another piece of candy in your mouth and allow it to dissolve *using your tongue and teeth.* Record the time (in seconds) it takes for the **candy shell** to dissolve.

Part C: Mystery Colors

Follow the steps to use chromatography to separate the pigments in black ink. You will need three black markers, three pieces of filter paper, a piece of pipe cleaner, and a small beaker of water.

16. Use the marker to draw a circle (about the size of a dime) in the center of a piece of filter paper.

17. Insert one end of a pipe cleaner into the center of your ink dot. Place the other end into the water in the beaker.

18. Allow time for the water to move up the pipe cleaner and separate the pigments in the ink.

19. Repeat the first three steps to test the other markers.

20. Observe colors that are present in each marker and record in Part C of the data sheet.

Part D: See the Light

21. Create four different mixtures by following the steps below.

    a. Mix 10 g of salt with 100 mL of water in a Ziploc bag. Label the bag as D-1.

    b. Mix 10 g of flour with 100 mL of water in a Ziploc bag. Label the bag as D-2.

    c. Mix 10 g of Kool-Aid powder with 100 mL of water in a Ziploc bag. Label the bag as D-3.

    d. Mix 10 g of dirt with 100 mL of water in a Ziploc bag. Label the bag as D-4.

22. Shine the light of a laser pointer through each bag. Record your observations of each mixture in Table 1.3.

23. Read the following information from the Columbia Encyclopedia, and then answer the questions on your Data Sheet.

> *One property of a colloid that distinguishes it from a true solution is that the particles in a colloid scatter light. If a beam of light passes through a colloid, the light is reflected or scattered by the particles in the colloid, and the path of the light can be observed. When a beam of light passes through a true solution, there is so little scattering of the light that the path of the light cannot be seen and the small amount of scattered light cannot be detected except by very sensitive instruments. The scattering of light by colloids, known as the Tyndall effect, was first explained by the British physicist John Tyndall.*

DATA SHEET

Name: _______________________________   Partner(s): _______________________________

TA: _______________________________   Section: _______________   Date: _______________

Part A: Hit the Trail!

1. What do you see in Bag A?

2. Would this mixture be classified as a heterogeneous or homogeneous mixture? Give a reason for your answer.

TABLE 1.1			
Material	**Mass (g)**	**Mass of Mixture (g)**	**% of Mixture**
Round final percentages to the nearest hundredth!			Total:

3. If everyone in the class had the same mixture, would they have the same results? Explain this in terms of how you know it's a mixture and not a compound.

Part B: Tasty Solutions

TABLE 1.2	
Piece of Candy	**Dissolving Time (s)**
First	
Second	
Third	

4. In your solution, what was the solute and the solvent?

    a. solute: **b.** solvent:

5. Explain the results of your experiment in terms of dissolving rate or the time it takes for a substance to dissolve.

6. Identify the solute(s) and solvent in each solution. **_Underline_** the solute, and **_circle_** the solvents. Remember that a **solute** dissolves in a **solvent!**

    a. ocean water: salt and water

    b. antifreeze: water and ethylene glycol

    c. soda pop: syrup, water, and CO_2 gas

    d. gold jewelry: gold and copper

    e. Kool-Aid: powder, sugar, and water

    f. lemonade: water, lemon juice, and sugar

    g. air: nitrogen, oxygen, and other gases

    h. sterling silver: silver and copper

7. What liquid is called the "universal solvent"?

8. Which would have the most **solute:** a glass of very sweet Kool-Aid or a glass of barely sweet Kool-Aid? Give a reason for your answer.

Part C: Mystery Colors

9. What happened to the black ink?

10. What colors did you observe for each pen?

    a. pen 1:   **b.** pen 2:   **c.** pen 3:

11. Identify the solute and solvent for this experiment.

    a. solute: **b.** solvent:

12. What do you think would happen if you used a permanent marker? Explain your answer.

Part D: See the Light

TABLE 1.3	
Bag	**Observations**
D-1 = Salt and Water	
D-2 = Flour and Water	
D-3 = Kool-Aid Powder and Water	
D-4 = Dirt and Water	

13. Which mixtures would be classified as colloids?   and

14. Which mixtures would be classified as solutions?   and

15. Name the solutes and solvents for the solutions.

    a. bag   → solute: solvent:

    b. bag   → solute: solvent:

16. Describe a situation in which you would observe the Tyndall Effect.

TA Signature	
Pre-Lab Assignment ________ Safety/Participation ________ Lab Write-Up ________	Ask your TA to review your work and sign your report. The TA will sign above once satisfied that the student has performed the entire procedure. The report will not be accepted or graded unless signed.

Density Exploration

Objectives

- Students will learn the significance of uncertainty in laboratory measurements.
- Students will calculate densities and use those densities to identify unknowns.

Safety Precautions and Hazards

- Chemical splash goggles and lab coats must be worn at all times.
- Nitrile gloves must be worn when handling chemicals.

PRE-LAB QUESTIONS

Name: _________________________________ Partner(s): _________________________________

TA: _________________________________ Section: _____________ Date: _____________

Answer the following questions using information obtained from lecture, the textbook, or lab manual. The responses will be checked at the start of your lab section for credit. Failure to complete may result in exclusion from participating in the lab.

Example: A student has a sample of aluminum that has a mass of 27 g and a volume of 10.0 mL. What is the density of aluminum?

Density = mass/volume

Density = 27 g/10.0 mL

Density = 2.7 g/mL

1. Volume measurements will be made using a 10-mL graduated cylinder, which has scale divisions every 0.1 mL. What is the estimated uncertainty in the volume measurements?

2. The volume of a metal cylinder was measured indirectly by water displacement three times. The following volume readings were recorded. Calculate the **average volume** of the metal cylinder.

Initial Volume (cm^3)	Final Volume (cm^3)
7.6	19.3
8.0	20.1
6.2	18.4

3. A loaf of bread has a mass of 500.0 g and volume of 2500 mL. What is the density of the bread?

BACKGROUND

Often in taking measurements the scale reading must be estimated because the quantity to be measured falls between two divisions on the instrument scale. Consider the portion of the thermometer shown in Figure 2.1. The level of the fluid (represented by the red region) falls between two scale divisions of the thermometer.

We know with certainty that the measurement is between 67°F and 68°F. Exactly where the reading lies in this range can only be estimated. One might guess that it is just above the halfway mark and record the measurement as 67.7°F. In recording measurements it is understood that the last digit is an estimate whereas all other digits are known with certainty (in this case the 67 is known with certainty).

In reporting measurements, it is common to indicate the uncertainty in the measurement. Uncertainty refers to the interval around the measurement with which the researcher is confident the measurement lies. The uncertainty is typically taken as one half the smallest

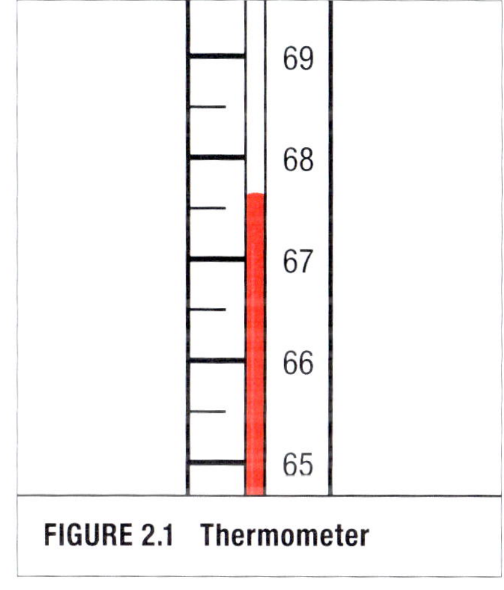

FIGURE 2.1   Thermometer

numerical subdivision on the measuring device. In this case, the uncertainty would be chosen to be 0.5°F and the measurement could be recorded as 67.7°F ± 0.5°F. This implies that the researcher estimated the reading to the tenths place but is confident the measurement lies between 67.2°F and 68.2°F. Although this uncertainty range may seem larger than necessary, since it is obvious the measurement is not larger than 68°F, this is a simple and common approach to handling uncertainties in measurements.

What Is Density?

It is defined as amount of ***anything*** per unit volume. Simply put, it is how much stuff can fit into a space. Chemists most often refer to mass density, which is found by dividing the mass of substance (g) by the volume of the substance (mL, cm^3, or cc). Density is dependent on the type of substance, not how much of it is present. If given a whole candy bar or half of the same candy bar, the density of the candy is the same. In this lab, you will explore different methods to obtain the density of substances. You will also be identifying substances based on their densities.

Once you determine a density experimentally, you will need to evaluate the precision and accuracy of your results. We shall use a calculation called *percent range* to express precision and a calculation called *percent error* to express accuracy. The necessary formula follows:

$$\% \text{ Error} = \frac{(\text{"True" Value} - \text{Average Experimental Value})}{(\text{"True" Value})} \times 100\%$$

All measurements must include the uncertainty of the piece of equipment.

Accepted Density Values

TABLE 2.1 Physical Constants of Inorganic Compounds (solids)

Substance	Density (g/cm³)
Lead	11.3
Gold	19.3
Copper	8.96
Aluminum	2.70
Iron	7.87
Zinc	7.14
Tin	7.265
Potassium Chloride	1.988

TABLE 2.2 Physical Constants of Organic Compounds (liquids)

Substance	Density (g/cm³)
Glycerol	1.261
Acetone	0.789
Glycerol	1.2613
2-Propanol (Isopropyl Alcohol)	0.7809
Octanol	0.830
Water	0.998
Cottonseed Oil	0.921

PROCEDURE

Part A: Density of Liquids

Pure Water

1. Obtain approximately 5 mL of the known liquid. Record the volume exactly in Table 2.3 on the Data Sheet.

2. Place a clean, dry, and empty 50- to 100-mL beaker onto a digital scale. To "zero" the digital scale with the graduated cylinder on it, press the "re-zero" or "tare" button. Make sure to always use the same scale for measurements.

3. Place the liquid into the dry beaker (from Step 2). Determine and record the mass of the liquid using the digital scale. Re-zero the scale.

4. Repeat the procedure, this time using approximately 8 mL of the liquid.

5. Using your mass and volume measurements, determine the density of the liquid for each liquid sample.

6. Calculate the average density value.

7. Calculate the percent error for your density values.

Different Liquid

8. Empty and dry the beaker.

9. Record the number (or name) of the new liquid.

10. Repeat the above steps using the new liquid.

11. Identify the liquid; compare your density results to those in Table 2.1 and Table 2.2.

Part B: Density of Solids

Irregular Solid

12. Obtain a solid unknown sample. Place a weigh boat on your balance. Hit "tare" or "zero" to remove the weight of the boat. Place the sample in the boat and find the mass.

13. Add approximately 50 mL of water to a 100-mL graduated cylinder. Read and record the **water volume.**

14. Carefully place your solid sample in the graduated cylinder with the water. Gently agitate the graduated cylinder to eliminate trapped air bubbles.

15. Read and record the **new water volume.** This is the volume of the water and the metal sample.

16. Calculate the volume and density of your sample.

17. Identify the solid; compare your density results to those in Table 2.1 and Table 2.2.

A Real Life Application of Density

Problem: Cans of Diet Coke and regular Coke both have the same volume and the same amount of liquid inside them. Predict what will happen when the cans are placed in water. Record your predictions on the Data Sheet. Try it!

18. Find the mass and volume of each type of Coke.

19. Look at the can to determine how many grams of sugar are present in the regular Coke.

20. Weigh one sugar cube. Determine how many sugar cubes are needed for one can of regular Coke.

21. Weigh out 0.100 g of Nutra Sweet (aspartame). Record mass.

22. Compare these amounts.

DATA SHEET

Name: ________________________________   Partner(s): ________________________________

TA: ____________________________________   Section: ______________   Date: ________________

Part A: Density of Liquids

Pure Water

Show calculations for all items marked with an asterisk (*).

TABLE 2.3 Known Liquid Sample		
Identity of Known Sample		
True Value of Density of Known Sample	________ g/mL	
	Trial 1	**Trial 2**
Mass of Liquid	________ ± ________ g	________ ± ________ g
Volume of Liquid	________ ± ________ g	________ ± ________ g
Density of Liquid*	________ g/mL	________ g/mL
Average Density of Liquid*	________ g/mL	
Percent Error*	________ %	

Work for density of liquid:

Work for average density of liquid:

Work for percent error:

Different Liquid

Show calculations for all items marked with an asterisk (*).

TABLE 2.4 Unknown Liquid Sample		
Number of Unknown		
	Trial 1	**Trial 2**
Mass of Liquid	_______ ± _______ g	_______ ± _______ g
Volume of Liquid	_______ ± _______ g	_______ ± _______ g
Density of Liquid*	_______ g/mL	_______ g/mL
Average Density of Liquid*	_______ g/mL	
Percent Error*	_______ %	

Work for density of liquid:

Work for average density of liquid:

Work for percent error:

Part B: Density of Solids

Irregular Solid

Show calculations for all items marked with an asterisk (*).

TABLE 2.5	
Unknown Sample Letter	
Mass of Sample	_________ ± _________ g
Initial Water Volume	_________ ± _________ mL
Final Water Volume	_________ ± _________ mL
Volume of Sample*	_________ mL
Density of Sample*	_________ g/mL
Identity of Unknown (from Table 2.1)	
Density of Unknown (from Table 2.1)	
Percent Error*	_________ %

Work for volume of sample:

Work for density of sample:

Work for percent error:

Questions

1. Explain why a highly precise measurement does not always mean a highly accurate one.

2. Why is the density of a substance independent of sample size?

3. Identify at least two possible sources of error in the procedure used to determine the density of your unknown liquid.

A Real Life Application of Density

1. Predict what will happen when a can of regular Coke is placed in the water container.

2. Predict what will happen when a can of Diet Coke is placed in the water container.

3. Describe what happened when you placed a can of regular Coke in the container of water.

4. Describe what happened when you placed a can of Diet Coke in the container of water.

TABLE 2.6	Regular Coke	Diet Coke
Mass of Each Can		
Volume of Liquid in Can		
Density of Each Can*		

Work for density of each can:

5. What ingredient is in the regular Coke and not in the Diet Coke, giving it more mass?

6. How many grams of sugar are in a can of regular Coke?

7. Mass of one sugar cube: _______ → (_______ sugar cubes/can)

8. Mass of Nutra Sweet: _______

9. How does the mass and density of regular Coke differ from the mass and density of Diet Coke? Explain why they differ.

TA Signature	
Pre-Lab Assignment _______ Safety/Participation _______ Lab Write-Up _______	Ask your TA to review your work and sign your report. The TA will sign above once satisfied that the student has performed the entire procedure. The report will not be accepted or graded unless signed.

Calories in Snack Foods

Objectives

- Students will demonstrate the ability to collecte data to determine heat energy.
- Students will be able to calculate the number of Calories in various snack foods.

Additional Reading and References

- TopHat: OpenStax General Chemistry 5.1–5.2

Safety Precautions and Hazards

- Chemical splash goggles and lab coat must be worn at all times.
- Long hair must be tied back to protect from flames.
- Extinguish all matches completely before disposing.

PRE-LAB QUESTIONS

Name: _________________________________   Partner(s): _________________________________

TA: ___________________________________   Section: _______________   Date: _______________

Answer the following questions using information obtained from lecture, the textbook, or lab manual. The responses will be checked at the start of your lab section for credit. Failure to complete may result in exclusion from participating in the lab.

All human activity requires "burning" food for energy. How much energy is released when food burns in the body? How is the caloric content of food determined? Let's investigate the caloric content of different snack foods such as nuts, marshmallows, and cheese puffs.

1. Identify:

    a. independent variable:

    b. dependent variable:

    c. control variables (at least two):

        i.

        ii.

2. Write your hypothesis here:

    The

    is related to the   .

BACKGROUND

In this lab we are going to experiment with various snack foods to determine the amount of energy released when the food is burned. This involves the scientific method and identifying variables involved in the process. Read the following definitions to help answer the Pre-Lab Questions.

Variable: Something that is changed. In scientific experiments there are two variables: *one that you control and one that is the result.*

1. **Independent Variable:** *"The Cause"*

   - The one thing that *you* changed in an experiment

   - This variable makes one test "independent" of another test.

   - On a graph, it is on the *x*-axis (along the horizontal side).

2. **Dependent Variable:** *"The Effect"*

   - The result of the experiment; *what* is measured

   - This "depends" on what you changed.

   - On a graph, it is on the *y*-axis (along the vertical side).

Control Variable: The things/conditions that remain unchanged in the experiment.

The *dependent variable* is related to the *independent variable.*

What does it mean to say that we burn food in our bodies? The digestion and metabolism of food converts the chemical constituents of food to carbon dioxide and water. This is the same overall reaction that occurs when organic molecules—such as carbohydrates, proteins, and fats—are burned in the presence of oxygen. The reaction of an organic compound with oxygen to produce carbon dioxide, water, and heat is called a *combustion reaction.* The chemical equation for the most important reaction in our metabolism, the combustion of glucose, is shown in Eq. (1).

$$C_6H_{12}O_6 + 6O_2 \rightarrow 6CO_2 + 6H_2O + \text{heat} \tag{1}$$

Within our bodies, the energy released by the combustion of food molecules is converted to heat energy (to maintain our constant body temperature), mechanical energy (to move our muscles), and electrical energy (for nerve transmission). The total amount of energy released by the digestion and metabolism of a particular food is referred to as its *caloric content* and is expressed in units of nutritional **Calories (note the uppercase C) which is the same as 1 kilocalorie.** The caloric content of most prepared foods is listed on their nutritional information labels.

Nutritionists and food scientists measure the caloric content of food by burning the food in a special device called a calorimeter. ***Calorimetry* is the measurement of the amount of heat energy produced in a reaction.** Calorimetry experiments are carried out by measuring the temperature change in **water** that is in contact with or surrounds the reactants and products. (The reactants and products together are referred to as the system, the water as the surroundings.)

Materials

- pop can
- balance
- aluminum foil
- paperclip (food holder)
- matches
- cheese puff
- nut
- thermometer
- graduated cylinder
- file or glass rod
- big marshmallow

PROCEDURE

Compare the number of calories and *Calories per gram in a nut, a cheese puff, and a marshmallow.*

1. Assemble the ring stand, clamp, pop can, and food holder. Suspend the pop can about 2.5 cm above the food being tested on the holder.

2. Place **40.0 mL** of water in the empty soda can, and suspend the pop can on the ring.

3. Place the thermometer in the pop can, being careful to not let it touch the bottom of the can. Record the temperature of the water in the can to the nearest **0.5°C** and record in Table 3.1 on the Data Sheet as initial water temperature.

4. Obtain three different types of snack food: a nut (your choice of type), a marshmallow, and a cheese puff.

5. Choose one type of snack food to begin the experiment. Place the snack food onto the paperclip. Place the paperclip on a small square of aluminum foil. Record the mass of the food, the food holder, and the foil together. Record its mass in Table 3.1 on the Data Sheet.

6. Place the food sample on the table away from the can. Set the food on fire. *Immediately* move the sample (on foil) under the can. Let the sample burn itself out completely. Relight it *immediately* if it goes out before all of the food is burnt to a black crisp.

7. After the food sample is completely burned, measure the temperature of the water again to the nearest **0.5°C** and record in Table 3.1 on the Data Sheet as final water temperature. **Be sure to watch the thermometer carefully to catch the highest temperature reached.**

8. Find the mass of the sample remaining in the "food holder" and on the foil and record in Table 3.1 on the Data Sheet as mass of sample after burning (ash weight).

9. Repeat procedure for the other snack food samples. You will need fresh water each time. You should have three total snack foods tested.

Before You Leave Lab

- Clean up by dumping water, wiping off can, putting any snack food burnt or not used into the trash, and returning equipment to proper place.

- Wipe down your lab station!

DATA SHEET

Name: _______________________________   Partner(s): _______________________________

TA: _______________________________   Section: ______________   Date: ______________

TABLE 3.1						
Food Sample	Mass before Burning (g) (food, holder, and foil)	Mass after Burning (g) (remaining food, holder, and foil)	Change in Mass (g)	Initial Temperature (°C)	Final Temperature (°C)	Change in Temperature (°C)
Nut						
Marshmallow						
Cheese Puff						

Calculations

1. Determine how much heat energy in *calories* is absorbed by the water for each snack food. You should be using

   $q = mc\Delta T$ (*m* is the mass of the water here)

   Hint: Water has a density of 1.00 g/mL, which means that 1 mL = 1 g of water. Use this to convert volume of water to mass of water.

   specific heat (c_{water}) of 1 cal/g°C
   or
   4.18 J/g°C

   a. nut (calories) q_{H_2O} =

   b. marshmallow (calories) q_{H_2O} =

   c. cheese puff (calories) q_{H_2O} =

2. Due to the law of conservation of energy, we know **heat lost = heat gained.** This means that

$$-q_{lost} = q_{gained}$$

So, if the water got hotter, where did the heat energy come from in this experiment?

3. If the water absorbed the heat, is that an endothermic or exothermic process?

4. What would be the sign of q for the heat of the snack food?

5. Calculate the calories per gram (cal/g) of snack food **given off** in this experiment. **Watch the sign on q.**

   a. nut (cal/g)

   b. marshmallow (cal/g)

   c. cheese puff (cal/g)

6. Now convert this to Cal/g.

   a. nut (Cal/g)

   b. marshmallow (Cal/g)

   c. cheese puff (Cal/g)

Questions

1. Which snack food has the higher energy content?

2. Comparing the nutritional labels (compare protein, fat, and sugars) of the two foods, hypothesize about why you think this is the case.

TA Signature	
Pre-Lab Assignment ________ Safety/Participation ________ Lab Write-Up ________	Ask your TA to review your work and sign your report. The TA will sign above once satisfied that the student has performed the entire procedure. The report will not be accepted or graded unless signed.

POST LAB

Name: ________________________________   Partner(s): ________________________________

TA: ________________________________   Section: ____________   Date: ______________

Answer the following questions after you have completed this lab.

A candy bar has a total mass of 75.0 grams. In a calorimetry experiment, a 1.0-g sample of this candy bar was burned in a calorimeter using 1000 g of water. The temperature of the water in contact with the burning candy bar was measured and found to increase from an initial temperature of 21.2°C to a final temperature of 24.3°C.

1. Calculate the amount of heat in calories released when the 1.0-g sample burned. Use water as your mass.

2. Convert the heat in calories to nutritional Calories and then calculate the energy content (fuel value) in Cal/g.

3. Calculate the **total** caloric content of the c.

Atomic Spectra, Discharge Simulation, and Flame Tests

Objectives

- Students will learn the fundamental origins of the line emission spectra displayed by different elements.

Additional Reading and References

- Dr. Ted M. Clark, the Ohio State University, Department of Chemistry and Biochemistry
- Top Hat: OpenStax General Chemistry 6.1–6.4

Safety Precautions and Hazards

- Chemical splash goggles and lab coat must be worn at all times.
- Nitrile gloves must be worn when handling chemicals.
- ***Do not*** touch solutions with your bare hands.
- Use caution with the burner.
- Metal salts go into the inorganic waste and not down the drain.
- Wood splints should be extinguished completely in sink and then disposed of in trash can.

PRE-LAB QUESTIONS

Name: ________________________________   Partners: ________________________________

TA: ________________________________   Section: ____________   Date: ________________

Answer the following questions using information obtained from lecture, the textbook, or lab manual. The responses will be checked at the start of your lab section for credit. Failure to complete may result in exclusion from participating in the lab.

1. A bookshelf analogy is often used to describe the electrons that surround the nucleus of an atom. Describe how this analogy relates to these electrons, referring to the textbook, if necessary.

2. Fill in the blanks below with higher or lower. When energy is absorbed by an atom, an electron may be excited from a ____________ energy level to a ____________ energy level. When light is emitted from an atom, an electron has transitioned from a ____________ energy level to a ____________ energy level.

BACKGROUND

When light is emitted from a single element (like gaseous hydrogen) and passed through a prism, only a small number of discrete lines are observed. Atomic models account for this phenomenon by proposing that an electron absorbs energy and moves to an excited state. When the electron subsequently moves to a lower energy level, a photon is emitted. The reason that discrete lines are observed, and not a continuous spectrum, is because the different energy level in the atom are quantized and have specific values.

When energy is added to atoms or ions of a particular element, electrons can absorb the energy and move to a higher energy level. When the electrons of these "excited" atoms or ions return to their normal ground state, energy is emitted.

As you have observed experimentally, the emission spectra for hydrogen and helium are different from each other. Also, the energy level diagrams may be different for different elements, and this means their emission spectra may be different. When you look at these excited atoms or ions with the naked eye, you see only a single color that is the "sum" of all the individual colors emitted. In this activity you will observe this phenomena for different metal salts that have been excited by heating in a flame.

Materials

- burner tongs
- wooden splints (pre-soaked)
- salts of various metals, such as
    - sodium
    - potassium
    - calcium
    - barium
    - strontium
    - lithium
    - copper

PROCEDURE

Part A: Computer Simulation— Neon Lights and Discharge Lamps

https://phet.colorado.edu/en/simulation/discharge-lamps

There are several variables you can manipulate within the simulation. See Figure 4.1.

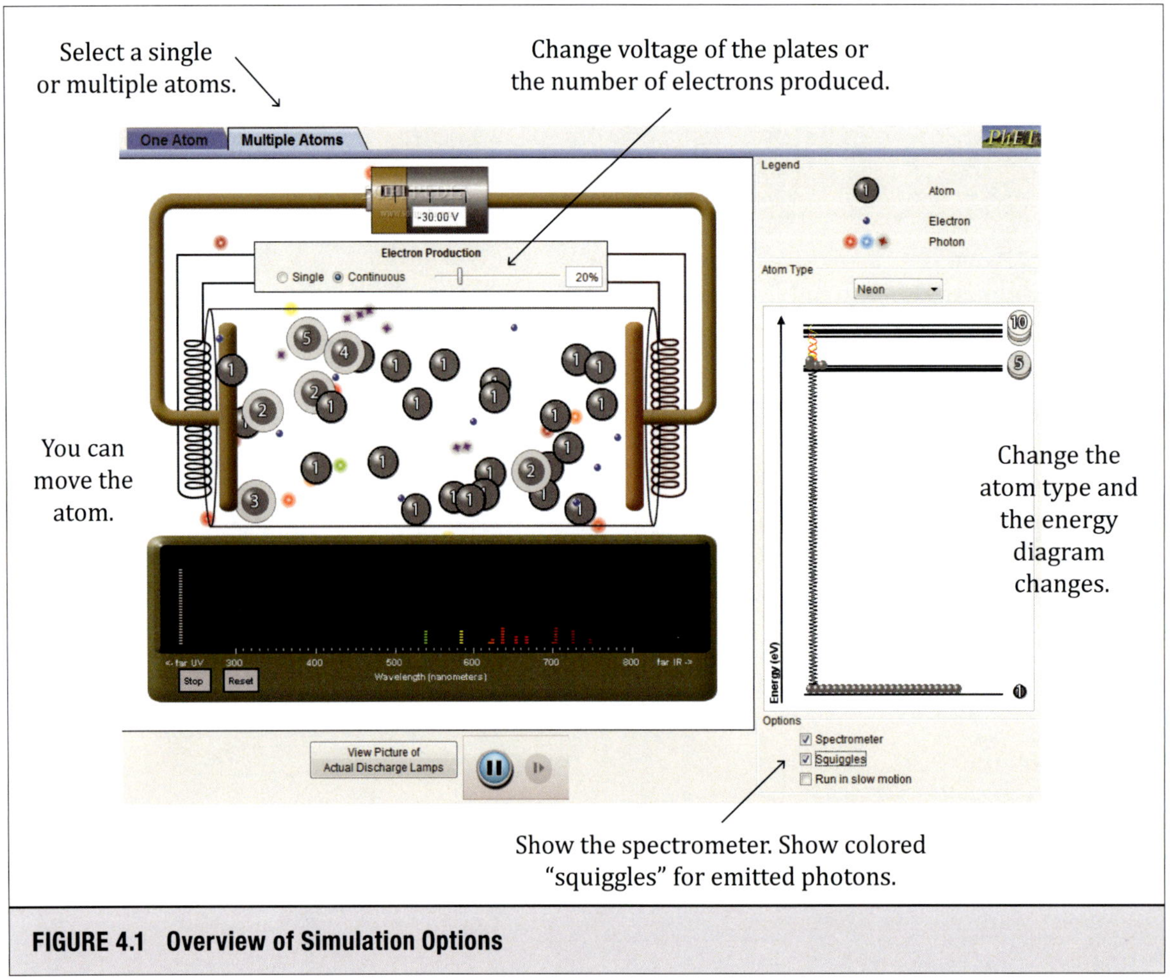

FIGURE 4.1   Overview of Simulation Options

1. To begin your investigation, sketch the electronic energy level diagram for the different atoms. To see the electronic energy levels, click on **"Atom Type"** and select the desired atom.

2. Sketch the electronic energy levels for each atom in the appropriate box on the Data Sheet.

   By having electrons strike the atom, you can cause it to emit photons, and the energy of the emitted photons is recorded by the **spectrometer.**

3. **Turn on your spectrometer.** You must turn on the spectrometer to record the emitted photons (right side of the screen, Figure 4.1).

4. The electron production can be **single** or **continuous** and the voltage may be adjusted (see the top of the screen, as shown in Figure 4.1). Please ***choose single and slow motion to begin.***

   You may turn it up to continuous once you know what's really happening.

5. In the "**Atom Type**" pull-down menu, you can configure (move) the energy levels and also add energy levels, or choose a particular element (like neon).

6. **Start producing electrons (FIRE!) and use the simulation to gather information.**

7. Select the "**One Atom**" tab, then answer the questions in Table 4.1.

8. Select the "**Multiple Atom**" tab, then answer the questions in Table 4.2.

Part B: Flame Tests of Metals

CAUTION	**Goggles must be worn throughout the experiment. Do not touch solutions with your bare hands. Use caution with the burner.**

9. Light and adjust your Bunsen burner; ask for assistance if you are unsure how to use the burner.

10. To do a flame test with each metal salt, get a wooden splint from the salt solution containing that metal and bring it into the hottest part of the flame. Observe the color produced from the metal salt. (Remember wood burns orange so if that is all you see, you may need to ask for assistance.) Once color is observed, extinguish splint in sink.

11. In Table 4.3 on the Data Sheet, record the **name** of each compound and the **overall color** of each when it is put in the flame.

Before You Leave Lab

- Metal salts go into the inorganic waste.
- Wood splints should be extinguished completely in sink and then disposed of in trash can.

DATA SHEET

Name: _________________________________   Partner(s): _______________________________

TA: ___________________________________   Section: _______________   Date: _______________

Part A: Computer Simulation— Neon Lights and Discharge Lamps

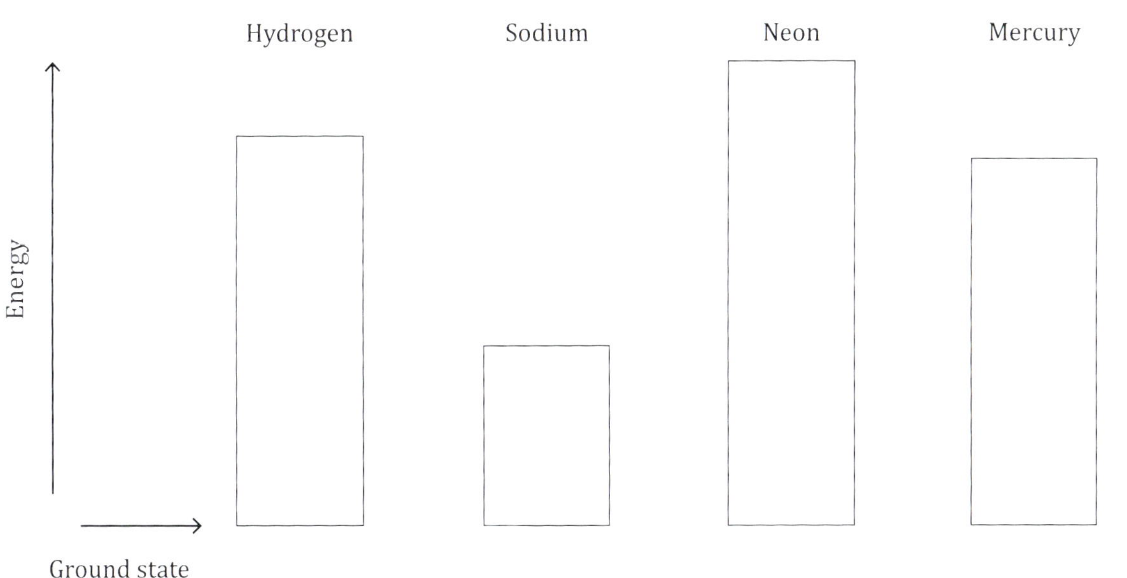

1. What is the electron configuration for each atom in its ground state?

a. hydrogen: **c.** neon:

b. sodium: **d.** mercury:

TABLE 4.1		
	Statement	**True or False**
1	The photons emitted from an excited atom are always a different color.	
2	The energy **emitted** from a discharge lamp is quantized. (***Hint:*** look at the spectrometer.)	
3	Photons are emitted as electrons move to a **higher** energy level.	
4	If the spacing between two electronic energy levels in atom A is larger than in atom B, then the **wavelength** of the light emitted by atom B will be longer. (You may want to use the configurable atom to answer this question.)	
5	If the spacing between two electronic energy levels in atom A is smaller than in atom B, then **fewer photons** will be emitted by atom B. (You may want to use the configurable atom to answer this question.)	
6	Electrons are destroyed when they strike an atom, causing an excited state.	

TABLE 4.2		
	Statement	**True or False**
1	Neutral atoms of different elements have different number of electrons. As the number of electrons increases, so too does the **number of spectral lines** in the visible region.	
2	For a collection of many atoms, some are in an excited state and others are in the ground state. At any point in time, most of the atoms have electrons in the ground state.	

Part B: Flame Tests of Metals

TABLE 4.3	
Metal Salt Solution Name/Formula	**Color of Flame**

TA Signature	
Pre-Lab Assignment _________ Safety/Participation _________ Lab Write-Up _________	Ask your TA to review your work and sign your report. The TA will sign above once satisfied that the student has performed the entire procedure. The report will not be accepted or graded unless signed.

POST-LAB

Name: ________________________________   Partner(s): ________________________________

TA: ________________________________   Section: ____________   Date: ____________

Answer the following questions after you have completed this lab.

1. List the elements tested in the flame test in order of increasing energy of the light emitted (based on the general color of the flame).

2. Is the emitted color of light a periodic trend? Is the trend (or lack of trend) consistent with what you observed in the discharge lamp simulation?

3. In your own words, explain what is happening to the electron during the flame test. Use the following terms: energy, ground state, excited state, wavelength, frequency, energy level, and photon.

4. Many fireworks contain metal salts like the ones you tested in this activity. Explain the chemistry that produces the colors.

Chemical Bonds, Molecular Models, and Shapes

Objectives

- Students will build and predict the molecular geometry of a variety of molecules.

Additional Reading and References

- Top Hat: OpenStax General Chemistry 7.2–7.3

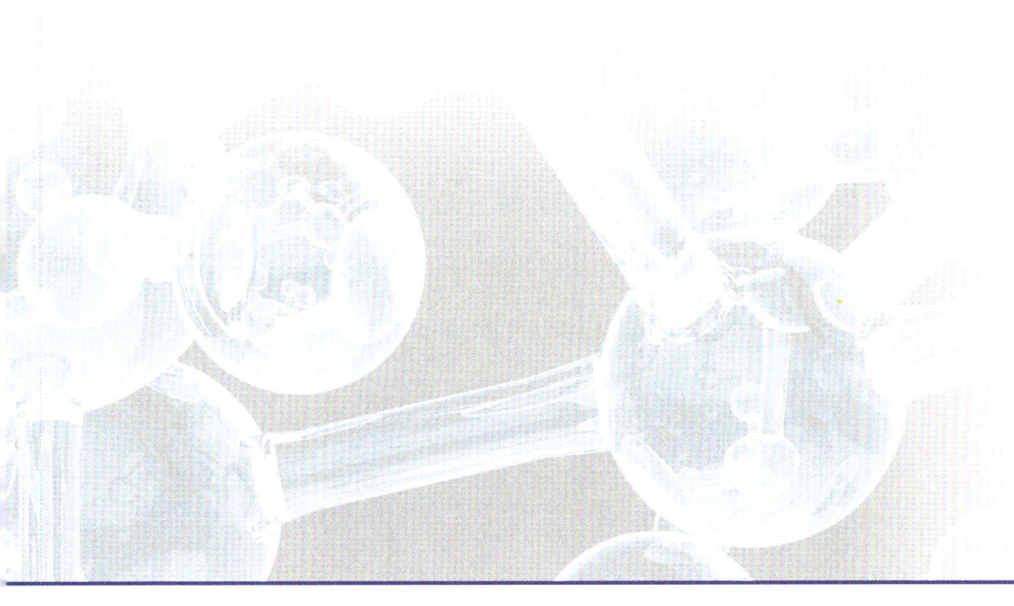

PRE-LAB QUESTIONS

Name: _________________________________   Partners: _________________________________

TA: _________________________________   Section: _____________   Date: _______________

Answer the following questions using information obtained from lecture, the textbook, or lab manual. The responses will be checked at the start of your lab section for credit. Failure to complete may result in exclusion from participating in the lab.

1. The valence electrons (i.e., the electrons in the highest energy shell of an atom) of a given atom are the electrons involved in the formation of chemical bonds. Referring to your textbook on the subject of valence electrons, determine how many valence electrons are associated with atoms of the following elements:

   a. hydrogen

   b. lithium

   c. sodium

   d. boron

   e. chlorine

   f. sulfur

   g. oxygen

   h. carbon

   i. nitrogen

   j. neon

   k. magnesium

2. In the formation of molecular compounds, many, but not all, atoms follow the "octet rule," which is the tendency for atoms to associate with eight electrons through the combination of non-bonding and bonding electron pairs. Referring to the textbook's description of Lewis dot structures, circle the bonding electron pairs in the following molecules:

a. CCl_4

$$: \overset{\cdot\cdot}{\underset{}{Cl}} :$$
$$: \overset{\cdot\cdot}{\underset{\cdot\cdot}{Cl}} : \overset{\cdot\cdot}{\underset{\cdot\cdot}{C}} : \overset{\cdot\cdot}{\underset{\cdot\cdot}{Cl}} :$$
$$: \underset{\cdot\cdot}{Cl} :$$

c. CO_2

$$: \overset{}{\underset{\cdot\cdot}{O}} :: C :: \overset{}{\underset{\cdot\cdot}{O}} :$$

b. NH_3

$$H$$
$$H : \overset{\cdot\cdot}{\underset{\cdot\cdot}{N}} : H$$

d. H_2O

$$H : \overset{\cdot\cdot}{\underset{\cdot\cdot}{O}} : H$$

BACKGROUND

It is important to understand the geometry of molecules—to "have a feel for" their shapes and sizes. This is because the physical properties and the chemical reactivities of molecules are related to their sizes and shapes. In this experiment you will construct models of several molecules and ions so that you can see (and feel) the spatial arrangements of the atoms.

You may need to consult your text or others before lab to determine how the atoms are connected to one another.

When predicting structures of molecules, it is often useful to draw first their Lewis dot structures. This will help visualize how the valence electrons are shared to form bonds. The following set of guidelines will work in most cases.

1. Count the number of **valence** electrons (Group number for A Group elements) for each atom and sum them. If the species of interest is a negative ion, add number of electrons equal to the negative charge; if positive, deduct.

2. Draw a skeletal structure joining atoms by single bonds, keeping in mind that:

    a. Hydrogens can form only **one** bond;

    b. Oxyanions usually have oxygens arranged around the central atom;

    c. Except for oxyanions and interhalogen compounds, halogens usually form only one bond;

    d. Carbons are often bonded together;

    e. The central atom is ***usually*** written first in the formula: NH_3, ClO_2^-, OF_2 (exceptions: H_2O, H_2S).

3. Deduct two valence electrons for each single bond written in Step 2.

4. Distribute the remainder of the valence electrons to attempt to give each outer atom an octet of electrons. If any electrons remain, add them to the central atom.

5. If there are too few electrons on the central atom, make multiple bonds by taking a lone pair from one outer atom and sharing it with the central atom to which it is connected.

6. Look at the number of bonds formed and number of lone pair electrons **versus** the number of electrons the atom had originally. Most atoms should follow the octet rule (8 total electrons based on the number of non-bonding electrons plus the number of bonding electrons), with H being the main exception.

Example CCl_4

1. Number Valence Electrons

Element	Number of Atoms	Number of Electrons	Total Electrons
C	1	4	4
Cl	4	7	**28**
			32
No Charge			**0**
		Total	32

2.

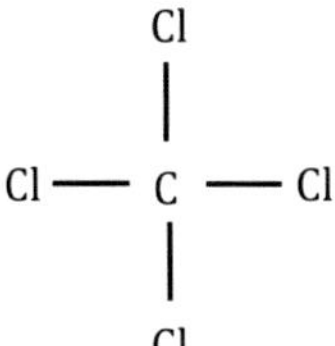

3. Four single bonds use eight electrons, so 24 e⁻ left.

4. C has an octet, and distributing 6e⁻/Cl also gives each chlorine atom an octet and uses all e⁻.

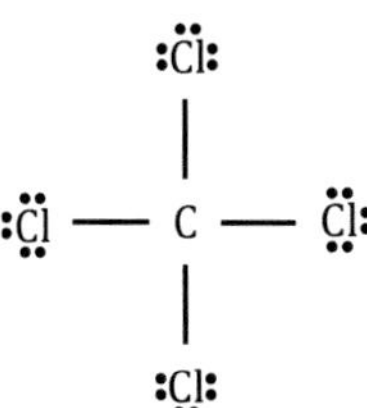

5. No multiple bonds needed to make up octets.

6. All atoms follow the octet rule, so this structure is reasonable.

Example PCl_3

1. Number Valence Electrons

Element	Number of Atoms	Number of Electrons	Total Electrons
P	1	5	5
Cl	3	7	**21**
			26
Charge			**0**
		Total	26

2.

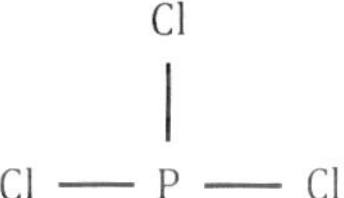

3. Three single bonds use six e⁻, so 20 e⁻ left.

4. Distributing 6e⁻/O also gives O atoms octets, and placing 2e⁻ on the P atom gives the central atom an octet and uses all e⁻.

5. No multiple bonds needed to make up octets.

After writing the Lewis dot structure, count the number of atoms (no matter whether singly or multiply bonded) and number of unshared (lone) electron **pairs** about the central atom. This approach is called the **VSEPR** (Valence Shell Electron Pair Repulsion) Model.

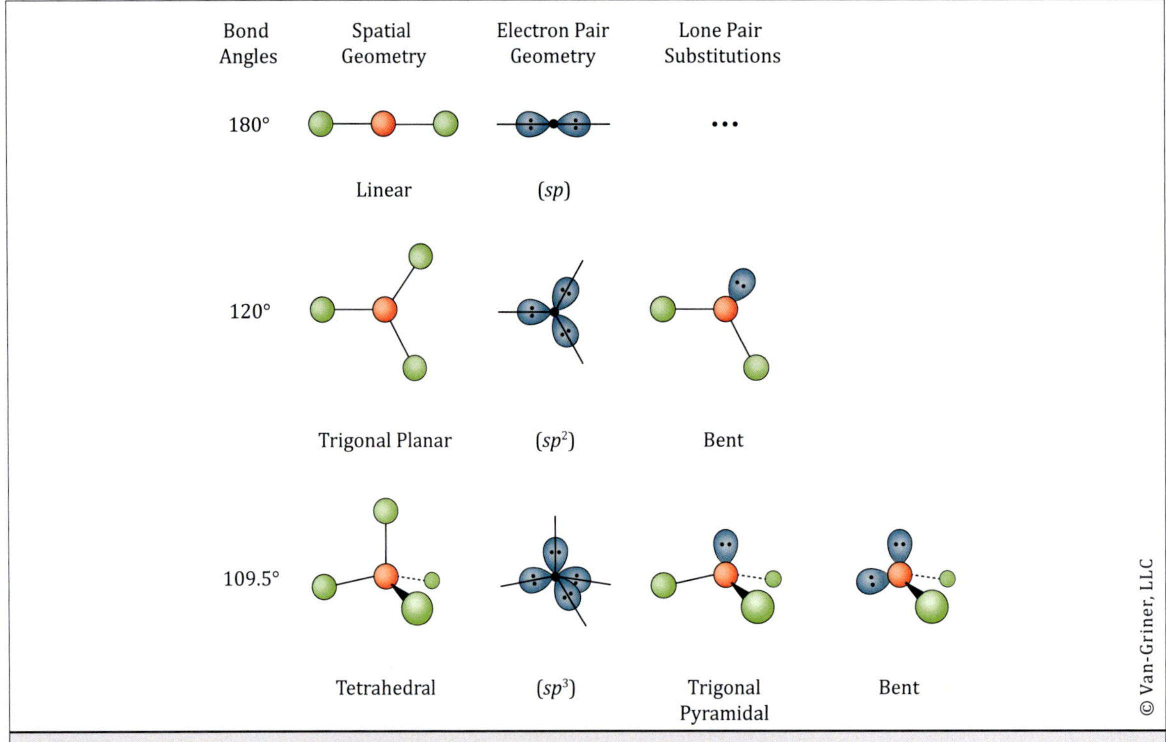

FIGURE 5.1   Summary of Geometry and Shapes Predicted by VSEPR Model

Molecular Geometry

TABLE 5.1 Molecular Geometry Chart

# of Electron Groups	Number of Lone Pairs	Electron Pair Arrangement	Molecular Geometry	Approximate Bond Angles
2	0	linear		180°
3	0	trigonal planar		120°
	1	bent		<120°
4	0	tetrahedral		109.5°
	1	trigonal pyramidal		<109.5° (107°)
	2	bent		<109.5° (105°)
5	0	trigonal bypyramidal		90°, 120°
	1	see-saw		<90°, <120°
	2	T-structure		<90°
	3	linear		180°
6	0	octahedral		90°, 90°
	1	square pyramidal		90°, <90°
	2	square planar		90°

Model Building Basics

Molecular model kits vary; therefore, your instructor will explain the particular models that you will use. The kit probably contains balls (used for atoms), sticks (used for single bonds and unshared electron pairs), and springs or curved sticks (used for double and triple bonds). Each stick or spring represents two electrons. Hydrogen atoms are usually represented by small, light-colored balls (yellow, white, or pale blue) that have only one hole. The color code for other atoms will vary. A common set of colors is shown in Table 5.2.

TABLE 5.2 Molecular Model Kits
Molecular model kits usually presume certain color conventions for atoms, as follows.

Atom	Color
Hydrogen	Yellow or White
Carbon	Black
Nitrogen	Blue
Oxygen	Red
Fluorine or Chlorine	Green or Purple
Sulfur	Yellow

Note: There is one disadvantage to using the colored balls provided in most model sets. They usually have only enough holes for the correct number of bond pairs, and thus you will not be able to see the unshared electron pairs. An alternate approach is to use balls with four holes in them for all atoms other than hydrogen so that the octet (four pairs of electrons) will always be visible.

PROCEDURE

Each pair of students should have a model set. First, get acquainted with the components of the set. Note the holes in the various colored balls and their positions. If there are two lengths of sticks, the short ones are for bonds involving hydrogen, and the longer ones are for any other single bond.

Using the procedure outlined below, build models for each of the molecules listed on Table 5.3. Then use information obtained from viewing the models to fill in the information in the last two columns of the table. You should take time to think about (and write down in words and a diagram) the shape of each molecule before proceeding to the next one.

6. Assemble the atoms required. (For example, to make the CH_4 molecule, you will need one black or blue sphere and four yellow ones.) Next, note the group in the periodic table to which each element belongs. The number of the group is also the number of outer electrons in an atom of the element.

7. To determine how many sticks (pairs of electrons) you will need, divide the total number of outer electrons by two. For example, H_2O has one outer electron from each hydrogen and six from oxygen, for a total of eight. Hence, you will need four sticks to represent all the electrons in H_2O. Two sticks represent bonds between H and O, and two sticks represent unshared electron pairs.

8. If there is only one atom of one element in the molecule and more than one atom of another element, the single atom usually goes in the center of the molecule. This is the case in CO_2, but there are a few exceptions to this rule (such as N_2O, which has the arrangement NNO).

9. With the collected parts, assemble the model in such a way that each atom except hydrogen has a share in an octet of electrons. If you do not appear to have enough sticks (electron pairs) to give each atom (except hydrogen) an octet, try sharing more electrons by forming double or triple bonds (replace straight sticks with curved sticks or springs).

DATA SHEET

Name: ________________________________   Partner(s): ________________________________

TA: ________________________________   Section: ____________   Date: ____________

TABLE 5.3				
Molecule	**Total e⁻ Pairs**	**Number of Lone Pairs on Central Atom**	**Draw Lewis Structure**	**Electron Pair Arrangement**
Methane CH_4				
Ammonia NH_3				
Water H_2O				
Fluorine F_2				
Oxygen O_2				
Nitrogen N_2				
Carbon Dioxide CO_2				
Phosphine PH_3				
Sulfur Difluoride SF_2				
Dichloroethylene $C_2H_2Cl_2$				
Hydrazine N_2H_4				

TA Signature

Pre-Lab Assignment ________
Safety/Participation ________
Lab Write-Up ________

Ask your TA to review your work and sign your report. The TA will sign above once satisfied that the student has performed the entire procedure. The report will not be accepted or graded unless signed.

POST LAB

Name: _________________________________   Partner(s): _________________________________

TA: _________________________________   Section: _____________   Date: _____________

Answer the following questions after you have completed this lab.

1. The tetrahedral shape is one of the most fundamental shapes in chemical compounds. How would you describe it in words to someone who has never seen it?

2. The octet rule appears to be a very important rule governing the structures of molecules. In light of your work with models, provide a simple explanation for the importance of eight electrons.

3. Explain in your own words why nonbonded electron pairs help determine the shapes of molecules.

4. Do all of the assigned molecules obey the octet rule? If not, why (or in what way) did the octet rule fail?

5. As a test of what you have learned, predict the shapes of

    a. NF_3 **b.** H_2S **c.** Cl_2O

6. Models do not necessarily have to be physical objects. They can be 2D drawings or even mental constructs. Cite one or more examples of such models encountered outside of chemistry. Can you think of models that are used in your own field of study or that you will use in your future career?

Intermolecular Forces

Objectives

- Classify the type of intermolecular forces (hydrogen bonding, dipole-dipole interaction, and London dispersion forces) present in covalent molecules.
- Infer the relative strength of intermolecular forces based on observations of physical properties.

Additional Reading and References

- Top Hat: OpenStax General Chemistry 10.1

Safety Precautions and Hazards

- Chemical splash goggles and lab coats must be worn at all times.
- Many liquids being used in this activity are highly volatile, meaning they evaporate quickly at room temperature. Keep all containers closed when they are not in use so that unnecessary vapors do not escape into the room.

PRE-LAB QUESTIONS

Name: _________________________________   Partner(s): _________________________________

TA: _________________________________   Section: _______________   Date: _______________

Answer the following questions using information obtained from lecture, the textbook, or lab manual. The responses will be checked at the start of your lab section for credit. Failure to complete may result in exclusion from participating in the lab.

1. Draw a structural formula for each compound (look them up). You will use these in **Station A.**

    a. ethanol C_2H_5OH

    b. glycerin $C_3H_5(OH)_3$

2. How can you determine by looking at the formula if a molecule can form hydrogen bonds?

    a. What is a hydrogen bond?

    b. What types of elements are involved with hydrogen bonding?

3. How many hydrogen bonds can form in each compound?

    a. ethanol:

    b. glycerin:

4. **a.** What are dispersion forces? (LDFs)

   b. What is another name for a dispersion force?

5. Draw a structural formula for each compound (look them up). You will use these in *Station B.*

   a. propane C_3H_8 **b.** hexane C_6H_{14} **c.** paraffin $C_{18}H_{38}$

6. Draw the structural formulas for each compound. You will use these in *Station 4.*

   a. C_3H_7OH **b.** H_2O **c.** CH_3OH

7. Define solubility.

PROCEDURE

Station 1: Comparing Liquids with Hydrogen Bonding

1. There are two flasks with different liquids: C_2H_5OH (ethanol) and $C_3H_5(OH)_3$ (glycerin). **Do not** remove the stoppers from the flasks. Test them by swirling the contents and comparing the time it takes for the fluid motion to stop and the difficulty you have in moving the fluid quickly. Also shake the flask and observe how long it takes for the bubbles to disappear. Record your observations in Table 6.1 on the Data Sheet, then answer the following questions.

   Identify the liquid in each flask and give your reason. Your reason should relate your observations to the number of hydrogen bonds. Describe how **viscosity** *(the resistance of a liquid to flow)* is related to intermolecular forces. *Would a viscous liquid (like molasses) have strong or weak intermolecular forces? Would the compound with the most hydrogen bonding possibilities have a low or high viscosity?*

Station 2: Comparing Molecules with Only London Dispersion Forces

2. You have a sample of C_3H_8 (propane), C_6H_{14} (hexane), and $C_{18}H_{38}$ (paraffin). Observe the samples and compare the strength of the dispersion forces between the molecules. Record your observations in Table 6.2 on the Data Sheet, then answer the following questions.

   a. Identify each of the samples and explain your reasoning by relating observations to intermolecular forces.

   b. In general you can say that as the _____ of the molecule increases, the dispersion forces will increase. *How does size (or length) of a compound affect London Dispersion Forces? Do shorter or longer molecules have greater London Dispersion Forces?*

   c. What causes dispersion forces?

Station 3: Surface Tension and Strength of van der Waals Forces

3. There are two test tubes with plastic pipettes at this station. Place one drop of each of the liquids on a piece of plastic: $C_{12}H_{26}$ (oil) and H_2O (water).

4. Look at each droplet from the side. In the space provided on the Data Sheet, draw a picture of the drops made by each liquid.

5. In the space provided on the Data Sheet, draw the structural formula for each of the liquids at this station.

6. Identify the types of intermolecular forces found for each of the liquids and relate these to the shape of the drop as seen from the side. Include the terms **cohesive** and **adhesive** forces properly in your explanations. *What forces (dipole-dipole, hydrogen bonds, or dispersion forces) are present in each sample?* Record your answers on the Data Sheet.

Station 4: Using Evaporation to Compare Attraction between Molecules

7. At the same time, place one drop of each of the three liquids on the counter and observe the time it takes each to evaporate. The liquids are water, C_3H_7OH (rubbing alcohol), and methanol (CH_3OH). Indicate the relative time it takes each to evaporate—fastest, medium, slowest. Record your observations in Table 6.3 on the Data Sheet, then answer the following question.

   Identify the liquids and give a reason for your choice. Your reason should relate your observations to the intermolecular forces. *What are the IMFs in each compound (dipole-dipole, hydrogen bonds, or dispersion forces)? Which compound has the strongest IMFs? Which compound has the weakest IMFs? Would a compound with strong IMFs evaporate quickly or slowly (how easy will it overcome the IMFs between molecules)?*

Station 5: Using Viscosity to Compare Attraction between Molecules

8. You will find three test tubes at this station. Each test tube is capped and has a small plastic bead in it. Each test tube has a different liquid in it. For each liquid, follow these steps:

9. Hold the tube upright until the bead is at the bottom.

10. Quickly turn the tube upside down so that the bead is at the top.

11. Measure the time recorded for the bead to fall to the other end of the tube and record in Table 6.4 on the Data Sheet.

12. Repeat two more times and determine the average time for each liquid. Record your observations in Table 6.4 on the Data Sheet.

13. Answer the following questions on your Data Sheet.

    a. Compare the forces of attraction between molecules for the three liquids. Which has the strongest forces? Which has the weakest forces? *Would a viscous compound have strong or weak IMFs?*

    b. Polymer chemists use a test just like the one at this station to determine the molecular weight of polymers they manufacture. **Using this test, which liquid (assume they are all nonpolar molecules) has the largest molecules?** *Explain by relating your observations to the IMFs present in the liquids. How do dispersion forces relate to length/ weight of a compound? Do shorter or longer molecules have greater dispersion forces? In the experiment, would a more viscous substance (longer time for beads to reach end of the tube) be long or short in length?*

Station 6: Intermolecular Forces/Evaporation Lab

14. Divide a section of the lab benchtop in half. Wipe one half with a moist paper towel while your partner wipes the other half with a wet wipe. Time how long it takes for each half to dry. Record your observations in Table 6.5 on the Data Sheet, then answer the following questions.

   a. How does the presence of intermolecular forces affect the rate of evaporation of various liquids? *Why would alcohol wipes evaporate more quickly than water? Which compound has more prevalent hydrogen bonding?*

   b. Certain liquids must be stored in air-tight containers because they can evaporate quickly. Gasoline is an excellent example of this concept. Infer the type of intermolecular forces present in gasoline.

Station 7: Intermolecular Forces and Solubility (DEMO)

Solubility is very important in the paint and coatings industry as the pigment (solute) must be soluble in the vehicle (solvent). Thus, watercolors contain different pigments than those found in oil-based paint. Examine the solubility of each solute in each solvent.

15. Add approximately 0.1 g of $CuCl_2$ (very small scoopful) to approximately 25 mL of water, ethanol, and hexane (each in a separate test tube). Stir and record your observations in Table 6.6 on the Data Sheet.

16. Add approximately 0.1 g of iodine (very small scoopful) to approximately 25 mL of water, ethanol, and hexane (each in a separate test tube). Stir and record your observations in Table 6.6 on the Data Sheet.

17. Answer the following questions on your Data Sheet.

   a. Classify the bond characters of $CuCl_2$ and iodine.

   b. Draw each molecule and classify the overall polarity of each molecule.

   c. Explain the experimental results. *How does the phrase "like dissolves like" relate to the polarity of a molecule?*

DATA SHEET

Name: _________________________________ Partner(s): _________________________________

TA: _________________________________ Section: _____________ Date: _____________

Station 1: Comparing Liquids with Hydrogen Bonding

TABLE 6.1		
Letter of Sample	Results of Swirling	Results of Shaking
A		
B		

1. Identify the liquid in each flask and give your reason. Your reason should relate your observations to the number of hydrogen bonds. Describe how **viscosity** *(the resistance of a liquid to flow)* is related to intermolecular forces. *Would a viscous liquid (like molasses) have strong or weak intermolecular forces? Would the compound with the most hydrogen bonding possibilities have a low or high viscosity?*

A:

 Reason:

B:

 Reason:

Station 2: Comparing Molecules with Only London Dispersion Forces

TABLE 6.2		
Letter of Sample	State (phase) of Sample	Relative Strength of Intermolecular Forces (high, med, low)
C		
D		
E		

2. Identify each of the samples and explain your reasoning by relating observations to intermolecular forces.

C:

Reason:

D:

Reason:

E:

Reason:

3. In general you can say that as the   of the molecule increases, the dispersion forces will increase. *How does size (or length) of a compound affect London Dispersion Forces? Do shorter or longer molecules have greater London Dispersion Forces?*

4. What causes dispersion forces?

Station 3: Surface Tension and Strength of van der Waals Forces

5. Look at each droplet from the side. Draw a picture of the drops made by each liquid.

a. $C_{12}H_{26}$ (oil)

b. H_2O (water)

6. Draw the structural formula for each of the liquids at this station.

    a. $C_{12}H_{26}$ (oil)   **b.** H_2O (water)

7. Identify the types of intermolecular forces found for each of the liquids on the Data Sheet and relate these to the shape of the drop as seen from the side. Include the terms **cohesive** and **adhesive** forces properly in your explanations. *What forces (dipole-dipole, hydrogen bonds, or dispersion forces) are present in each sample?*

Station 4: Using Evaporation to Compare Attraction between Molecules

TABLE 6.3	
Letter of Sample	**Relative Ranking of Rate of Evaporation**
F	
G	
H	

8. Identify the liquids and give a reason for your choice. Your reason should relate your observations to the intermolecular forces. What are the IMFs in each compound *(dipole-dipole, hydrogen bonds, or dispersion forces)? Which compound has the strongest IMFs? Which compound has the weakest IMFs? Would a compound with strong IMFs evaporate quickly or slowly (how easy will it overcome the IMFs between molecules)?*

F:

  Reason:

G:

Reason:

H:

Reason:

Station 5: Using Viscosity to Compare Attraction between Molecules

TABLE 6.4				
Letter of Sample	Trial 1	Trial 2	Trial 3	Average Time
K				
L				
M				

9. Compare the forces of attraction between molecules for the three liquids. Which has the strongest forces? Which has the weakest forces? *Would a viscous compound have strong or weak IMFs?*

10. Polymer chemists use a test just like the one at this station to determine the molecular weight of polymers they manufacture. **Using this test, which liquid (assume they are all nonpolar molecules) has the largest molecules?** *Explain by relating your observations to the IMFs present in the liquids. How do dispersion forces relate to length/weight of a compound? Do shorter or longer molecules have greater dispersion forces? In the experiment, would a more viscous substance (longer time for beads to reach end of the tube) be long or short in length?*

Station 6: Intermolecular Forces/Evaporation Lab

TABLE 6.5		
Benchtop Swipe Test	**Time**	**Observations**
Damp Paper Towel		
Wet Wipe		

11. How does the presence of intermolecular forces affect the rate of evaporation of various liquids? *Why would alcohol wipes evaporate more quickly than water? Which compound has more prevalent hydrogen bonding?*

12. Certain liquids must be stored in air-tight containers because they can evaporate quickly. Gasoline is an excellent example of this concept. Infer the type of intermolecular forces present in gasoline.

Station 7: Intermolecular Forces and Solubility (DEMO)

TABLE 6.6		
Solvent	**$CuCl_2$ Solubility (Y/N)**	**Iodine Solubility (Y/N)**
Water		
Ethanol		
Hexane		

13. Classify the bond characters of $CuCl_2$ and iodine. **(ionic or covalent)**

    a. $CuCl_2$ = 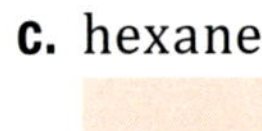   **b.** I_2 =

14. Draw each molecule and classify the overall polarity of each molecule.

    a. water   **b.** ethanol (C_2H_5OH)   **c.** hexane

15. Explain the experimental results. *How does the phrase "like dissolves like" relate to the polarity of a molecule?*

TA Signature	
Pre-Lab Assignment ________ Safety/Participation ________ Lab Write-Up ________	Ask your TA to review your work and sign your report. The TA will sign above once satisfied that the student has performed the entire procedure. The report will not be accepted or graded unless signed.

Exploring the MOLE

Objectives

- Students will investigate and understand that chemical quantities are based on molar relationships.

Additional Reading and References

- Top Hat: OpenStax General Chemistry 3.1

Safety Precautions and Hazards

- Chemical splash goggles must be worn at all times.
- Do not eat materials unless instructed to do so by the TA.

PRE-LAB QUESTIONS

Name: _______________________________________ Partner(s): _______________________________________

TA: _______________________________________ Section: _______________ Date: _______________

Answer the following questions using information obtained from lecture, the textbook, or lab manual. The responses will be checked at the start of your lab section for credit. Failure to complete may result in exclusion from participating in the lab.

1. You are given 18.6 g of Mg. Do you have a mole of magnesium? How do you know?

2. How many moles of magnesium are in 3.01×10^{22} atoms of magnesium? moles

3. Write the chemical formula for calcium phosphate.

4. How many formula units (same as molecules) are in 12.5 moles of calcium phosphate?

5. What is the percent of oxygen in calcium phosphate? %

6. You have 25.6 g of carbon dioxide.

    a. What is the molar mass of carbon dioxide? M

    b. How many molecules of carbon dioxide are in your sample? molecules

PROCEDURE

Please work with your peer groups on the following stations activity. For all stations, please clean up after observing/using the materials at each station. For the calculations, please show all work or receive **no credit!**

Station A: Aluminum—Moles and Atoms

1. The grey substance in the container is aluminum. Answer the questions on your Data Sheet.

Station B: Baking Soda/Vinegar—Molar Volume

2. Add approximately 10–15 mL of vinegar to an Erlenmeyer flask.

3. Add two spatulas of sodium hydrogen carbonate to a balloon.

4. Without spilling the solid into the flask, seal the flask with the balloon.

5. Weigh flask and balloon together. Record the mass in Table 7.1 on the Data Sheet.

6. Invert balloon so that the solid empties into the flask.

7. Release the gas collected by removing the balloon from the flask.

8. Record the mass of flask and empty balloon together.

9. Answer the questions on your Data Sheet.

Station C: Copper(II) Sulfate— Mass, Moles, Formula Units

10. The blue substance in the beaker is copper(II) sulfate. Answer the questions on your Data Sheet.

Station D: Snickers—Mass, Moles, Molecules

11. Read the label on the Snickers Bar, then answer the questions on your Data Sheet

Station E: Chalk—Mass, Moles, Formula Units, Atoms

12. Record the mass of a piece of chalk in Table 7.2 on the Data Sheet. ***Note:*** Chalk is calcium carbonate ($CaCO_3$).

13. Write all of the names of your group on a piece of paper. Record the new mass of the chalk.

14. Answer the questions on your Data Sheet.

Station F: Gum—Mass, Moles, Molecules

15. Place the wrapper or a piece of paper on the balance and press the "tare" button.

16. Record the mass of an unwrapped piece of gum in Table 7.3 on the Data Sheet.

17. Chew the piece of gum until all the sugar ($C_{12}H_{22}O_{11}$) is dissolved in your mouth. You may wish to return to this station later.

18. Remass the piece of gum on the piece of paper and record its mass.

19. Answer the questions on your Data Sheet.

Station G: Iron Nail—Mass, Moles, Atoms

20. Answer the questions on your Data Sheet.

Station H: Oreos—Percent Composition

21. Record the mass of a plain "tuxedo" cookie in Table 7.4 on the Data Sheet.

22. Carefully open the cookie, and scrape off the cream.

23. Find the mass of the cream and the mass of the brown cookie.

24. Record the mass of a double-stuffed "tuxedo" cookie.

25. Carefully open the cookie, and scrape off the cream.

26. Find the mass of the cream and the mass of the brown cookie.

27. Answer the questions on your Data Sheet.

DATA SHEET

Name: _______________________________________   Partner(s): _______________________________________

TA: _______________________________________   Section: _______________   Date: _______________

Station A: Aluminum—Moles and Atoms

1. Is this sample one mole of aluminum? (Y/N)   Explain your choice.

2. There are approximately 6.02×10^{23} atoms of aluminum in the sample. (Y/N) Explain your choice.

Station B: Baking Soda/Vinegar—Molar Volume

TABLE 7.1	Mass (g)
Flask and Balloon	
Flask and Empty Balloon	

3. Calculate the mass of the gas (CO_2) released.   g

4. Determine the number of moles of the gas (CO_2) released.   moles

5. Determine the moles of oxygen atoms released into the air.   moles

6. Assuming the room is at STP (22.4 Liters/mole), calculate the number of Liters of gas (CO_2) released.   Liters

Station C: Copper(II) Sulfate—Mass, Moles, Formula Units

7. What is the chemical formula of copper(II) sulfate?

8. What is the mass of this sample?   g

9. What is the molar mass of copper(II) sulfate?   g/mole

10. How many moles of copper(II) sulfate are there?   moles

11. Since copper(II) sulfate is ionic, it has formula units rather than molecules. Using your calculated number of moles, how many formula units/molecules are present in the sample?   formula units

Station D: Snickers—Mass, Moles, Molecules

12. How many grams of sugar (sucrose, $C_{12}H_{22}O_{11}$) are there in one bar?   g

13. How many moles of sucrose are there?   moles

14. How many molecules of sucrose are there?   molecules

Station E: Chalk—Mass, Moles, Formula Units, Atoms

TABLE 7.2	
	Mass (g)
Chalk	
Chalk after Writing	

15. Calculate the number of grams of chalk you used.   g

16. How many moles of chalk did you use? ***Reminder:*** Chalk is calcium carbonate ($CaCO_3$).
     moles

17. How many formula units (molecules) of calcium carbonate did you use?   units

18. If there are three atoms of oxygen in each formula unit ($CaCO_3$), how many atoms of oxygen did you use?   atoms

Station F: Gum—Mass, Moles, Molecules

TABLE 7.3		
		Mass (g)
Gum		
Gum after Chewing		
Sugar		

19. What is the mass of sugar in the gum?   g

20. How many moles of sugar are there in the piece of gum?   moles

21. How many molecules of sugar did you remove from the gum through chewing?   molecules

22. What percent of the unchewed piece of gum was sugar?   %

Station G: Iron Nail—Mass, Moles, Atoms

23. What is the symbol for the element iron?

24. What is the mass of one mole of iron?   g/mole

25. What is the mass of this nail?   g

26. How many moles of iron are in this nail?   moles

27. How many atoms of iron are in this nail?   atoms

Station H: Oreos—Percent Composition

TABLE 7.4		Mass (g)
Plain "Tuxedo" Cookie	Whole Cookie	
	Cream	
	Brown Cookies	
Double-stuffed "Tuxedo" Cookie	Whole Cookie	
	Cream	
	Brown Cookies	

28. Calculate the percent cream in the plain "tuxedo" cookie. ________ %

29. Calculate the percent cream in the double-stuffed "tuxedo" cookie. ________ %

30. Does the double-stuffed "tuxedo" cookie really have double the cream?

TA Signature	
Pre-Lab Assignment ________ Safety/Participation ________ Lab Write-Up ________	Ask your TA to review your work and sign your report. The TA will sign above once satisfied that the student has performed the entire procedure. The report will not be accepted or graded unless signed.

Types of Reactions

Objectives

- Students will classify the various types of chemical reactions observed in the experiment.

Additional Reading and References

- Top Hat: OpenStax General Chemistry 4.1–4.2

Safety Precautions and Hazards

- When heating a test tube, aim the mouth of the test tube *away* from yourself and others.

- Lab coat, goggles, and gloves must be worn at all times.

- ***Do not*** look directly at burning magnesium. The intense light may damage your eyes.

- Hydrochloric acid can cause burns to the skin and damage the eyes.

- Follow all disposal directions as given by your instructor.

PRE-LAB QUESTIONS

Name: _________________________________ Partner(s): _________________________________

TA: _________________________________ Section: _____________ Date: _____________

Answer the following questions using information obtained from lecture, the textbook, or lab manual. The responses will be checked at the start of your lab section for credit. Failure to complete may result in exclusion from participating in the lab.

For each of the following experimental procedures and observations:

    a. Write a balanced chemical equation.

    b. Name the product(s).

    c. Identify the type of reaction.

1. A student heats barium metal over a flame, and it begins to react with oxygen gas in the air. A white, crystalline solid begins to form on the barium.

    a. $\quad Ba + \quad O_2 \rightarrow \quad BaO$

    b.

    c.

2. A chemist combines lead(II) nitrate and potassium iodide solutions (both of which are clear liquids) and mixes them. A yellow precipitate appears in the final solution. The remaining liquid is an ionic solution. Solid precipitates containing lead are always yellow.

    a. $\quad Pb(NO_3)_2 + \quad KI \rightarrow \quad PbI_2 + \quad KNO_3$

    b.

    c.

3. Magnesium metal is placed in sulfuric acid. The solution begins to bubble. The remaining liquid is an ionic solution.

    a. $\quad Mg + \quad H_2SO_4 \rightarrow \quad MgSO_4 + \quad H_2$

    b.

    c.

4. Water is placed in an electrolysis machine. The water placed in the machine begins to bubble. The gas is collected in two separate containers. It is noticed that when a glowing splint is placed in one gas, it flames up. In the other gas, a glowing splint produces a small popping sound signifying a tiny explosion.

    a. $\quad H_2O \rightarrow \quad H_2 + \quad O_2$

    b.

    c.

BACKGROUND

There are an infinite number of chemical reactions. Chemists have divided these into broad classifications based on certain criteria. The most important classifications are **combination, decomposition, single replacement, double replacement, combustion, acid-base, and redox.** Note that some reactions will fall into more than one classification. For example, all single replacement reactions are also redox reactions.

Reactions are described with chemical equations. A chemical equation is the symbolic representation of the chemicals involved in the reaction. These chemical equations can be used to describe both physical processes and chemical reactions. All chemical equations consist of reactants, the starting materials, products, the ending materials, the state of matter that the materials are in, and a reaction arrow representing that a reaction has occurred.

> **reactants → products**

The states of matter are symbolized by subscripts which follow each chemical. The most common states of matter are solid, liquid, gas, and aqueous. The first three should be self-explanatory. The final, aqueous, occurs when a substance is dissolved in water. Gases and ions are commonly found in an aqueous state. The symbols are as follows: **solid(s), liquid(l), gas(g),** and **aqueous(aq).** Here are two examples of chemical equations:

> $H_2O(s) \rightarrow H_2O(l)$
>
> $CH_4(g) + O_2(g) \rightarrow CO_2(g) + H_2O(g)$

The first is the physical process of ice melting, and the second is the combustion of methane gas. There is one other aspect necessary in writing a proper chemical equation; the equation must have the same number of each type atom on each side of the reaction arrow. This is the **law of conservation of matter;** in other words, you are not allowed to create nor destroy matter. The first of the two above equations is fine, two hydrogens on each side and one oxygen on each side. The second equation needs to be balanced. This can be accomplished by placing a two in front of both the $O_{2(g)}$ and the $H_2O_{(g)}$.

> $CH_4(g) + 2O_2(g) \rightarrow CO_2(g) + 2H_2O(g)$

On each side we find one carbon, four hydrogens, and four oxygens. This equation obeys the conservation of matter and is said to be balanced.

Combination or synthesis reactions: two or more reactants combine to form one product. The general equation is

> **A + B → C**

A specific example is the tarnishing of a silver tea set:

> $Ag(s) + O_2(g) \rightarrow Ag_2O(s)$

Decomposition reactions: one reactant breaks to form two or more products. The general equation is

> **A → B + C**

A specific example is the electrolysis of water. Electricity causes this reaction:

$$2H_2O(l) \xrightarrow{V} 2H_2(g) + O_2(g)$$

Single replacement reactions: one chemical replaces another in a compound. The general equation is

$$A + BX \rightarrow B + AX$$

A specific example is zinc metal replacing iron in the iron(III) oxide compound. Ships use this reaction to keep their hulls from rusting:

$$3Zn(s) + Fe_2O_3(s) \rightarrow 2Fe(s) + 3ZnO(s)$$

Double replacement reactions: chemicals in each of two compounds switch compounds. The general equation is

$$AB + CD \rightarrow AD + CB$$

A specific example allows for the removal of toxic barium from a water source by adding a compound containing sulfate:

$$BaCl_2(aq) + Na_2SO_4(aq) \rightarrow BaSO_4(s) + 2NaCl(aq)$$

Combustion reactions: a hydrocarbon burns in the presence of oxygen gas to produce carbon dioxide and water. The general equation is

$$C_xH_y + O_2 \rightarrow CO_2 + H_2O$$

A specific example is the burning of candle wax:

$$C_{25}H_{52}(s) + O_2(g) \rightarrow CO_2(g) + H_2O(l)$$

PROCEDURE

For each reaction you perform, make observations of both the reactants the products formed. If no observable reaction occurred, write *no reaction or NR*. **This lab will use a flaming or smoldering piece of wood as a qualitative test for either $CO_{2(g)}$ or $O_{2(g)}$.** *If the fire goes out, you have CO_2 being generated; if the fire flames up, you have O_2 being generated.*

Reaction A

1. Add about 1 mL (1 mL ≈ width of your pinky finger) of calcium chloride solution, $CaCl_2$, to a clean test tube. Record observations in Table 8.1 on the Data Sheet. Next, add about 1 mL of sodium carbonate solution, Na_2CO_3, to the same test tube. Record your observations after the reaction.

Reaction B

2. Stand a clean test tube in the test tube rack and add 1–2 mL of hydrochloric acid.

Hydrochloric acid can cause burns to the skin and damage the eyes.

3. Record observations about HCl and Mg before the reaction in Table 8.1 on the Data Sheet.

4. Carefully drop a small strip of magnesium into test tube. While the reaction is still occurring, use a test tube holder to hold an empty test tube over the top of the first test tube for ~30 seconds.

5. Meanwhile, another group member should light the burner and use it to ignite a wood splint.

6. At the end of 30 seconds, hold the inverted test tube (keeping it inverted) while the other group member places the burning wood splint into the mouth of the inverted test tube to test for the presence of hydrogen gas.

7. Record your observations, including the results of the hydrogen test. ***Keep the burner lit for Reaction C.***

Reaction C

8. Place two spatula tips full of copper(II) carbonate, $CuCO_3$, in a clean test tube. Note the appearance of the sample in Table 8.1 on the Data Sheet.

9. Using a test tube holder, heat the $CuCO_3$ strongly in the burner flame for 2–3 minutes. ***Aim the mouth of the test tube away from yourself and others.*** After heating, another group member should ignite a wood splint and quickly place the burning splint into the test tube to test for the presence of CO_2 gas.

10. Record your observations, including the results of the CO_2 test.

Reaction D

11. Add about 5 mL of 1 M copper(II) sulfate, $CuSO_4$, solution to a clean, dry test tube. Place a small amount of zinc metal in the solution. Note the appearance of the solution and the zinc before and after the reaction. Record your observations in Table 8.1 on the Data Sheet. Allow the reaction to sit until after Reaction E. Come back to observe and make sure the reaction is complete.

Reaction E

12. Add four drops of 1 M NaCl, sodium chloride, to each of three separate, clean test tubes. **Number the test tubes 1, 2, and 3.**

13. Add four drops of 0.10 M $Cu(NO_3)_2$, copper(II)nitrate, to test tube 1.

14. Add four drops of 0.10 M $AgNO_3$, silver nitrate to test tube 2.

15. Add four drops of 0.10 M $Fe(NO_3)_3$, iron(III) nitrate, to test tube 3.

16. Agitate each test tube to mix the contents and record the observations for each mixture in Table 8.1 on the Data Sheet.

Reaction F: DEMO by TA

17. Place a watch glass next to the burner. Examine a piece of magnesium ribbon and record your observations in Table 8.1 on the Data Sheet.

18. Using crucible tongs, hold the ribbon in the burner flame until the magnesium starts to burn. ***Do not look directly at the flame.*** Hold the burning magnesium directly over the watch glass.

19. When the ribbon stops burning, place the remains on the watch glass. Examine this product thoroughly and record your observations.

Before You Leave Lab

- Clean your test tubes and lab station.
- Place liquids in appropriate waste containers.
- Solids should ***not*** go down the drain!

DATA SHEET

Name: _________________________________ Partner(s): _________________________________

TA: _________________________________ Section: _____________ Date: _____________

TABLE 8.1		
Reaction	**Before Reaction**	**After Reaction**
A	$CaCl_2$ Na_2CO_3	
B	HCl Mg	
C	$CuCO_3$	
D	$CuSO_4$ Zn	
E	NaCl $Cu(NO_3)_2$ $AgNO_3$ $Fe(NO_3)_3$	**Products** Test Tube 1: Test Tube 2: Test Tube 3:
F	Mg	

The following equations represent each actual reaction that occurred for **Reactions A through F**. First ***balance*** each equation. Then ***classify*** the reaction as one (or more) of the five general types of reactions—synthesis **(S)**, decomposition **(D)**, single replacement **(SR)**, double replacement **(DR)**, or combustion **(C)**.

Complete and Balance	**Type of Reaction**

A. $CaCl_2(aq) +$ $Na_2CO_3(aq) \rightarrow$ $NaCl(aq) +$ $CaCO_3(s)$

B. $Mg(s) +$ $HCl(aq) \rightarrow$ $MgCl_2(aq) +$ $H_2(g)$

C. $CuCO_3(s) \xrightarrow{\Delta}$ $CuO(s) +$ $CO_2(g)$

D. $Zn(s) +$ $CuSO_4(aq) \rightarrow$ $ZnSO_4(aq) +$ $Cu(s)$

E. $Cu(NO_3)_2(aq) +$ $NaCl(aq) \rightarrow$ $CuCl_2(aq) +$ $NaNO_3(aq)$
      $AgNO_3(aq) +$ $NaCl(aq) \rightarrow$ $AgCl(s) +$ $NaNO_3(aq)$
      $Fe(NO_3)_3(aq) +$ $NaCl(aq) \rightarrow$ $FeCl_3(aq) +$ $NaNO_3(aq)$

F. Magnesium reacts with oxygen to produce magnesium oxide.

TA Signature

Pre-Lab Assignment ________ Safety/Participation ________ Lab Write-Up ________	Ask your TA to review your work and sign your report. The TA will sign above once satisfied that the student has performed the entire procedure. The report will not be accepted or graded unless signed.

POST-LAB

Name: ___________________________________   Partner(s): ___________________________________

TA: _____________________________________   Section: ______________   Date: _______________

Answer the following questions after you have completed this lab.

1. What signs did you observe that indicated a chemical reaction was taking place? Be specific.

2. A combustion reaction was taking place each time you used the Bunsen burner to burn methane gas (CH_4). Write a balanced chemical equation for the combustion of methane.

3. Let's say you were doing an experiment on two different chemicals in two different test tubes. One chemical is giving off oxygen gas or hydrogen gas, and the other is giving off carbon dioxide gas. What one test could you do in each test tube that would show which is giving off the oxygen or hydrogen and which is giving off the carbon dioxide.

Mass of a Reaction Product (Stoichiometry)

Objective

- Students will use the principles of stoichiometry to determine the theoretical yield of a simple reaction, measure the actual yield, and calculate the percent yield.

Additional Reading and References

- Top Hat: OpenStax General Chemistry 4.1–4.3
- Developed in conjunction with Courtney Lantz, Pacifica High School (retired).

Safety Precautions and Hazards

- Chemical splash goggles and lab coat must be worn at all times.
- Nitrile gloves must be worn when handling chemicals.

PRE-LAB QUESTIONS

Name: ____   Partner(s): ____

TA: ____   Section: ________________   Date: ________________

Answer the following questions using information obtained from lecture, the textbook, or lab manual. The responses will be checked at the start of your lab section for credit. Failure to complete may result in exclusion from participating in the lab.

In this experiment you will react a measured amount of **sodium carbonate with an excess of hydrochloric acid (that is, more than enough to use up). The products of the reaction are sodium chloride, water, and carbon dioxide (a gas).** The carbon dioxide gas will bubble out of the solution and be lost. Measuring the decrease in mass gives the amount of carbon dioxide produced. Using the principles of stoichiometry you can then calculate the mass of carbon dioxide that should have been formed, and determine the percent yield of the experiment.

1. Write a balanced equation for the reaction that will take place in the lab.

2. What is the molar mass of carbon dioxide?

3. What is the molar mass of sodium carbonate?

4. Suppose you tare your balance to return it to read 0.000 g. You place a small plastic cup containing a white powder on the balance. You then add a small beaker holding two pipettes filled with acid. The total mass of this system is 159.621 g. You take the materials off of the balance and add enough acid to react completely with the powder. You use all of the acid from one pipette and only half of the other pipette. You make sure that the gas produced has left the cup. You then return the cup and the beaker with both pipettes to the balance. You find that the new mass of the system is 158.341 g. What is the mass of gas produced from the reaction?

BACKGROUND

A balanced chemical equation includes a great deal of useful information. Not only does it tell you in a concise manner what the reactants and products are, but it also tells the relative amounts of each substance.

In this experiment you will react a measured amount of sodium carbonate with an excess of hydrochloric acid (that is, more than enough to use up). The products of the reaction are sodium chloride, water, and carbon dioxide (a gas). The carbon dioxide gas will bubble out of the solution and be lost. Measuring the decrease in mass gives the amount of carbon dioxide produced. Using the principles of stoichiometry, you can then calculate the mass of carbon dioxide that should have been formed, and determine the percent yield of the experiment. Of course, if everything works exactly perfectly, the percent yield should be 100%—that is, the mass that is predicted to be formed should be the same as what is actually made.

Materials

- electronic balance
- 2 thin Beral pipettes
- 1 plastic cup
- plastic wrap
- 2 50-mL or 100-mL beakers

PROCEDURE

1. Put on your goggles. Take care in using the 3 M hydrochloric acid as it is corrosive. If you spill it on the lab table, you can neutralize it promptly with dilute sodium bicarbonate solution. If you spill some on your hand, rinse thoroughly and promptly with water in the sink.

2. Use the electronic balance to measure the mass of the plastic cup. First, use the "tare" button to make certain the balance reads "0.000 g" at the start, then place the cup on the center of the balance pan. Record the mass of the empty cup on the data sheet.

3. **Add something close to 0.50 g (between 0.45 g and 0.55 g) of sodium carbonate to the cup.** Record the exact mass of the cup with sodium carbonate (if you are using a triple beam balance), or of the added sodium carbonate (if you are using a digital scale). Be careful not to spill any sodium carbonate on the balance pan.

4. Wrap a small piece (a 4- or 5-inch square) of plastic wrap over the top of the cup.

5. Using a very sharp pencil or the tip of a scissors, poke three small holes in the plastic wrap that are just big enough to admit the tip of a Beral pipette. Distribute the holes so that they are equally spaced around the plastic on top of the cup about ¼ inch from the edge and about ¼ inch from each other.

6. Fill two pipettes completely with 3 M HCl solution and place them bulb down (tip up) in a beaker. Use special care when handing and transporting the hydrochloric acid.

Hydrochloric acid is corrosive and causes severe skin burns and eye damage.

7. Reset the balance to zero (0.000 g) and find the mass of all the equipment you have used so far: the beaker with the two filled pipettes and the condiment cup with sodium carbonate in it and the plastic wrap over it. Record this mass in Table 9.1. Remove set up from balance and return to lab station.

8. Begin the reaction: carefully insert the top of one of the Beral pipettes through one of the holes of the plastic wrap and lower it about halfway into the cup of sodium carbonate. Add the acid one drop at a time to the sodium carbonate. Add the acid slowly and wait for the bubbling to slow down before adding more acid. After adding about 10 drops of acid, gently swirl the solution to mix it. Continue to add acid and swirl gently until the reaction stops. You may have to use acid from the second pipette. Make certain that there are no tiny pieces of unreacted solid in the condiment cup.

9. Record your observations in Table 9.1 on the Data Sheet.

10. Add two more drops of acid to make certain that all of the sodium carbonate has reacted.

11. Carefully remove the system from the balance. Remove the plastic wrap, but do not discard. Waft the gas that was produced in the reaction. Continue wafting for 10–15 seconds directly above the mouth of the cup. *Replace* the plastic wrap.

12. Reset the balance to zero and again find the mass of all the pieces of equipment you have used (listed in Step 7).

13. Place both pipettes with any remaining acid back into their beaker and return them to the reagent table where you found them.

14. Slowly spill the reaction solution down the drain. Rinse it down with water. Thoroughly rinse out the condiment cup, dry it, and repeat for Trial 2.

DATA SHEET

Name: _________________________________   Partner(s): _________________________________

TA: ___________________________________   Section: ______________   Date: ______________

TABLE 9.1				
	Materials	**Mass (g)**	**Mass (g)**	**Average (g)**
Before Reaction	Empty Cup			
	Cup and Sodium Carbonate			
	Mass of Sodium Carbonate in Cup			
After Reaction	Cup, Sodium Carbonate, Plastic Wrap, and Two Pipettes			
	Cup, Reaction Products, Plastic Wrap, and Two Pipettes			

Observations:

TA Signature	
Pre-Lab Assignment _________ Safety/Participation _________ Lab Write-Up _________	Ask your TA to review your work and sign your report. The TA will sign above once satisfied that the student has performed the entire procedure. The report will not be accepted or graded unless signed.

POST LAB

Name: ________________________________   Partner(s): ________________________________

TA: ________________________________   Section: ____________   Date: ____________

Answer the following questions after you have completed this lab.

1. What is the actual yield of carbon dioxide produced? (Show your work.)

2. The word equation for the reaction is:

   Solid sodium carbonate reacts with hydrochloric acid solution to produce water, sodium chloride in solution, and carbon dioxide gas.

   Write and balance the chemical equation.

3. Use the periodic table to find and record the molar mass of each of the following two reagents to at least four significant figures:

   a. sodium carbonate: 1 mol = _______ g   **b.** carbon dioxide: 1 mol = _______ g

4. Determine how many grams of carbon dioxide you would have produced (if everything in this experiment worked perfectly) **given the mass of sodium carbonate that you added to the cup.** Show your work and give your answer to the correct number of significant figures.

Theoretical yield of carbon dioxide: _______ g

5. Determine the percent yield of carbon dioxide. Show your work and use a reasonable number of significant figures.

Percent yield: _______ %

6. Name and state the scientific law that requires your percent yield to be 100% if the entire reaction was perfect, and there were no errors in your procedure or measures.

Limiting Reactants and Percent Yield

Objectives

- Students will determine the limiting reagent of a reaction.
- Students will calculate theoretical, actual, and percent yield of a reaction.

Additional Reading and References

- TopHat: OpenStax General Chemistry 4.3–4.4

Safety Precautions and Hazards

- Chemical splash goggles and lab coat must be worn at all times.
- Nitrile gloves must be worn when handling chemicals.
- Copper(II) chloride is an irritant and moderately toxic. Avoid inhaling the powder and ingestion.
- Wash hands thoroughly after completing the lab.
- The reaction is exothermic and produces a lot of heat. Follow procedure carefully.
- ***Do not*** look into the beaker directly. Observe the reaction from the side.

PRE-LAB QUESTIONS

Name: _______________________________________   Partner(s): _______________________________________

TA: _______________________________________   Section: _______________   Date: _______________

Answer the following questions using information obtained from lecture, the textbook, or lab manual. The responses will be checked at the start of your lab section for credit. Failure to complete may result in exclusion from participating in the lab.

1. Write the balanced reaction of Al with $CuCl_2$.

2. What is the limiting reactant if 0.5 g Al (26.98 g/mol) is reacted with 3.5 g $CuCl_2$? $CuCl_2$ is a dihydrate with a molecular weight of 170.45 g/mol.

3. What is the theoretical yield of copper produced by this reaction?

BACKGROUND

During a chemical reaction when two substances react, often times one reactant will be consumed before the other. The substance that is consumed first is called the limiting reactant. This is the substance that controls or limits the amount of products formed.

Think about it in terms of making grilled cheese. If to make one grilled cheese, it takes two pieces of bread and one piece of cheese, and you have four pieces of bread and one piece of cheese, how many sandwiches can you make? The answer is one and therefore the cheese would be considered the limiting reactant. The two other slices of bread would be left unused and considered the excess reactant.

The limiting reactant for a specific reaction can be determined by calculating the amount of product that each reactant can produce. The reactant that forms the *least* amount of product will be the limiting reactant. The amount of product formed by the limiting reactant is the theoretical yield of the reaction.

The theoretical yield of a chemical reaction is the maximum amount of product that can be formed if the reaction proceeds perfectly. However, not all reactions proceed perfectly. Sometimes not all of the limiting reactant is used up, or some of the product can be lost during collection. The amount recovered from the reaction is known as the actual yield. The ratio of the actual yield to the theoretical yield is known as the percent yield and can be calculated using the formula below.

Percent Yield = (Actual Yield/Theoretical Yield) × 100

Materials

- copper(II) chloride
- aluminum foil
- stirring rod
- 2 100-mL beakers
- graduated cylinder
- safety glasses
- spatula or scoopula
- balance
- filter paper
- funnel
- watch glass
- ring stand
- iron ring

PROCEDURE

1. Weigh approximately 3.5 g of $CuCl_2$ and place them in the 100-mL beaker. Record the exact mass in Table 10.1 on the Data Sheet. Note the appearance of the $CuCl_2$ crystals in the observation section.

Copper(II) chloride causes skin irritation and serious eye damage. It is very toxic to aquatic life.

2. Add 50 mL of distilled water to the beaker containing the $CuCl_2$ crystals. Stir the solution until all of the $CuCl_2$ crystals have dissolved. Record any observations in the observation section.

3. Weigh out approximately 0.2 g of aluminum foil. Record the exact mass in Table 10.1.

4. Loosely crumple the piece of aluminum foil into a ball. Carefully place it into the $CuCl_2$ solution. Stir the reaction occasionally. Record your observations of the resulting reaction.

5. Set up a ring stand with an iron ring securely attached to it. Place a funnel in the iron ring.

6. Obtain a piece of filter paper and determine its mass. Record the mass in Table 10.1.

7. Fold a piece of filter paper in half and then in half again. Pull one side out and place it in the funnel.

8. Pour the reaction mixture into the funnel to separate the solid from the liquid. Make sure all of the solid is collected in the filter paper. Use a distilled water bottle to rinse the solid from the reaction beaker if necessary.

9. Pour the liquid collected into the waste beaker.

10. Allow the filter paper to dry in a low temperature oven for about 15 minutes.

11. Obtain the mass of the filter paper and solid. Record the mass in Table 10.1.

12. Place the solid in the solid waste beaker and throw out the filter paper.

DATA SHEET

Name: ____   Partner(s): ____

TA: ____   Section: _______________   Date: _______________

TABLE 10.1	
Mass of $CuCl_2$ (g)	
Mass of Aluminum Foil (g)	
Mass of Filter Paper (g)	
Mass of Filter Paper and Solid Product (g)	

TABLE 10.2   Observations
$CuCl_2$ crystals
$CuCl_2$ solution
Chemical reaction between aluminum foil and CuC_{l2}
Solution after reaction is complete

TA Signature

Pre-Lab Assignment _________
Safety/Participation _________
Lab Write-Up _________

Ask your TA to review your work and sign your report. The TA will sign above once satisfied that the student has performed the entire procedure. The report will not be accepted or graded unless signed.

POST LAB

Name: _______________________________________   Partner(s): _________________________________

TA: ____   Section: _______________   Date: _______________

Answer the following questions after you have completed this lab.

1. Calculate the amount of copper obtained by the reaction.

2. Calculate the theoretical yield of Cu. What was your limiting reagent?

4. What was the color of the solution after the reaction was complete? What does this tell you about the reactants?

5. Determine the percent yield of copper.

Salt Water Concentration

Objectives

- Students will prepare a solution and calculate the concentration of the solution in various ways.
- Students will design a method to determine the concentration of a new solution.

Additional Reading and References

- Top Hat: OpenStax General Chemistry 3.3–3.4

PRE-LAB QUESTIONS

Name: _______________________________________   Partner(s): _______________________________________

TA: _______________________________________   Section: _______________   Date: _______________

Answer the following questions using information obtained from lecture, the textbook, or lab manual. The responses will be checked at the start of your lab section for credit. Failure to complete may result in exclusion from participating in the lab.

The following procedures were followed in the lab. The data was recorded in Table 11.1.

1. Weigh a clean dry 100-mL beaker.

2. Add solid NaCl to approximately the 5-mL mark on the beaker (estimate if no mark).

3. Weigh the beaker again.

4. Add water to about the 40-mL mark and stir to dissolve the salt. Add more water until you have 50 mL of solution total.

5. Weigh again.

6. Pour into large graduated cylinder and record exact volume.

7. Use the data from Table 11.1 to fill out Table 11.2.

TABLE 11.1 Salt Concentration Lab	
Mass of Beaker (g)	100.00 g
Mass of Beaker and Salt (g)	107.60 g
Mass of Beaker, Salt, and Water (g)	157.80 g
Volume of Solution (mL)	50.5 mL

TABLE 11.2 Concentration Calculations	
Mass of Salt Used (g)	
Mass of Water (g)	
Mass of the Solution (g)	
Density (g/mL) of Solution Using D = m/v	
Molar Mass of NaCl	
Molar Mass of Water	
Moles of NaCl	
Moles of Water	

PROCEDURE

Part A

1. Weigh a clean dry 100-mL beaker. Record the mass in Table 11.3 on the Data Sheet.

2. Add solid NaCl to approximately the 5-mL mark on the beaker (estimate if no mark). Weigh the beaker again.

3. Add water to about the 40-mL mark and stir to dissolve the salt.

4. Add more water until you have 50 mL of solution total. Weigh again.

5. Record the total volume of the solution, using a graduated cylinder.

Part B

6. Pour out all but a small amount of your solution. Reweigh it. Record the mass in Table 11.4 on the Data Sheet.

7. Add enough water to make 50 mL of solution again. Weigh again.

8. Record the total volume of the solution, using a graduated cylinder.

9. Recalculate all information from Part A but not necessarily in that order.

10. Show all work with all units clearly labeled.

DATA SHEET

Name: ______________________________________   Partner(s): ______________________________________

TA: ______________________________________   Section: ______________   Date: ______________

Part A

TABLE 11.3	
Mass of Beaker (g)	
Mass of Beaker and Salt (g)	
Mass of Beaker, Salt, and Water (g)	
Volume of Solution (mL)	

1. Calculate the percent composition by mass.

    a. Calculate the mass of the solute.   g

    b. Calculate the mass of the solvent.   g

    c. Calculate the percent by mass of the NaCl in the solution.   %

2. Mole fractions tell how many moles of solute are present for every mole of total solution.

    a. Calculate the moles of NaCl solute.   moles

    b. Calculate the moles of water.   moles

    c. Calculate the mole fractions of your solution.   moles

3. Calculate the molarity of the solution.

4. Calculate the molality of the solution.

5. Calculate the density of the solution.   g/mL

Part B

TABLE 11.4	
Mass of Solution after Pouring Out (g)	
Mass of Solution after Adding Water (g)	
Volume of Solution (mL)	

6. Calculate the percent composition by mass.

    a. Calculate the mass of the solute.   g

    b. Calculate the mass of the solvent.   g

c. Calculate the percent by mass of the NaCl in the solution.   %

7. Mole fractions tell how many moles of solute are present for every mole of total solution.

a. Calculate the moles of NaCl solute.   moles

b. Calculate the moles of water.   moles

c. Calculate the mole fractions of your solution.   moles

8. Calculate the molarity of the solution.

9. Calculate the molality of the solution.

10. Calculate the density of the solution.   g/mL

<table>
<tr><td colspan="2">TA Signature</td></tr>
<tr><td>Pre-Lab Assignment ________
Safety/Participation ________
Lab Write-Up ________</td><td>Ask your TA to review your work and sign your report. The TA will sign above once satisfied that the student has performed the entire procedure. The report will not be accepted or graded unless signed.</td></tr>
</table>

Properties of Acids and Bases

Objectives

- Students will be able to classify substances as acid, base, or neutral based on the properties of the substance.

Additional Reading and References

- Top Hat: OpenStax General Chemistry 14.1, 14.2, and 14.7

Safety Precautions and Hazards

- Chemical splash goggles and lab coat must be worn at all times.
- Nitrile gloves must be worn when handling chemicals.

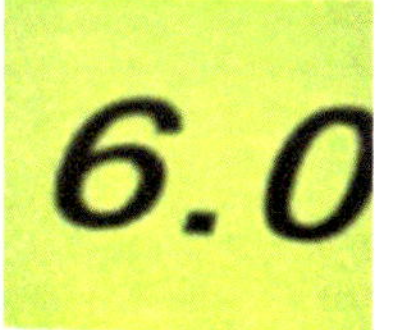
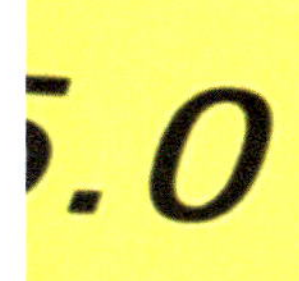

PRE-LAB QUESTIONS

Name: _______________________________   Partners: _______________________________

TA: _______________________________   Section: ____________   Date: ________________

Answer the following questions using information obtained from lecture, the textbook, or lab manual. The responses will be checked at the start of your lab section for credit. Failure to complete may result in exclusion from participating in the lab.

1. In the following acid-base equilibria, identify and label the acid and base among the reactants and the corresponding acid and base conjugates on the products side.

$$H_2O(l) + NH_3(aq) \rightleftharpoons OH^-(aq) + NH_4^+(aq)$$

$$HCl(aq) + H_2O(l) \rightleftharpoons H_3O^+(aq) + Cl^-(aq)$$

$$HCH_3CO_2(aq) + H_2O(l) \rightleftharpoons CH_3CO_2^-(aq) + H_3O^+(aq)$$

$$HCl(aq) + NH_3(aq) \rightleftharpoons NH_4^+(aq) + Cl^-(aq)$$

2. The relative strength of a series of acids can be determined by comparing the pH ($pH = -\log_{10}[H_3O^+]$) of solutions prepared from each of the acids. For a hypothetical acid, HA, the stronger the acid, the further to the right the following equilibrium lies:

$$HA(aq) + H_2O(l) \rightleftharpoons H_3O^+(aq) + A^-(aq)$$

Thus, stronger acids at a certain concentration will tend to produce more H_3O^+ ions than weaker acids in the same concentration. If a series of acids, A–E, produce pH values when prepared in identical concentrations of A: 3.5; B: 4.6; C: 3.1; D: 5.2; E: 3.8, arrange these acids in order of increasing acid strength:

______ <   ______ <   ______ <   ______ <   ______

BACKGROUND

Using the Brønsted concept of acids and bases, we have learned to consider acids as proton donors and bases as proton acceptors. We have also learned to think of an acid and its conjugate base (the acid minus its proton) or of a base and its conjugate acid (the base with an added proton). Some examples are shown in Table 12.1.

The last row of the table shows that water is amphoteric; i.e., it may act as an acid or as a base, depending upon the other chemical species in the environment. For example, in the reaction

$$HCl + H_2O \rightarrow H_3O^+ + Cl^-$$

water is acting as a base, receiving a proton from the strong acid hydrogen chloride. However, in the reaction

$$H_2O + NH_2^- \rightarrow NH_3 + OH^-$$

water is acting as an acid, donating a proton to the strong base, amide ion, NH_2^-.

TABLE 12.1 Examples of Acids, Bases, and Their Conjugates

Acid	Conjugate Base	Base	Conjugate Acid
HCN	CN^-	CN^-	HCN
NH_4^+	NH_3	NH_3	NH_4^+
H_2O	OH^-	H_2O	H_3O^+

Since all chemical reactions may be regarded as equilibrium processes if the reactants and products remain in the same environment, an acid-base reaction can be viewed in terms of the equilibrium established between the original acid and base and the conjugate base and conjugate acid formed. The equilibrium

$$HCN + H_2O \rightleftharpoons H_3O^+ + CN^-$$

is a typical example of an acid and a base reacting to give a new acid and a new base. The position of equilibrium always favors the weaker acid and weaker base since the stronger acid and stronger base have a greater tendency to react than do the weaker acid and base.

One way to measure the relative strengths of a series of acids is to allow each acid to react with the same base and to measure the position of equilibrium for each reaction. For example, in the series

$$HCl + HOH \rightleftharpoons H_3O^+ + Cl^-$$

$$CH_3CO_2H + HOH \rightleftharpoons H_3O^+ + CH_3CO_2^-$$

$$H_2PO_4^- + HOH \rightleftharpoons H_3O^+ + HPO_4^{2-}$$

$$NH_4^+ + HOH \rightleftharpoons H_3O^+ + NH_3$$

where water is the base in each reaction, one would discover that the position of equilibrium in the first process is far to the right, whereas that in the last process is far to the left. The positions of equilibrium for the second and third members of the series show a progression between that of the first and last members. Consequently, one could state that the order of acid strength in the series is $HCl > CH_3CO_2H > H_2PO_4^- > NH_4^+$. The position of equilibrium is often measured by electrical conductivity or by determining the voltage of a cell in which the solution under study serves as one electrode compartment and a standard hydrochloric acid solution serves as the other electrode compartment. The voltage is related to the position of equilibrium.

Indicators, which are themselves weak acids or bases, are sometimes used in studying acid-base equilibria. The usefulness of an indicator depends upon the fact that the acid form of the indicator, HInd, has a different color from that of the conjugate base form of the indicator, Ind$^-$, as shown in Table 12.2.

TABLE 12.2 Indicator Colors		
Indicator	**HInd Form**	**Ind$^-$ Form**
Phenolphthalein	Colorless	Pink
Litmus	Pink	Blue
Methyl Orange	Pink	Orange

The color of a solution containing an indicator reflects whether HInd or Ind$^-$ is present in higher concentration.

Since most indicators are weak acids (or weak bases), it should be possible to arrange a series of indicators in order of decreasing acid strength by studying the colors of their solutions with some acids of known strength. If the indicator is in its acid form (which we know from the color) when in a solution with another acid, HA, the indicator has been protonated by HA, therefore HA is "better" at donating protons than the indicator: HA is a stronger acid. If the indicator is in its base form when in a solution with HA, the indicator has donated a proton to A$^-$: the indicator is a stronger acid than HA. Perhaps this will be made clear from a consideration of

$$\textbf{HInd} + \textbf{A}^- \rightleftharpoons \textbf{HA} + \textbf{Ind}^-$$

Let us raise the question: "What statement can be made about the relative strengths of HInd and HA if the position of equilibrium is to the left or to the right?" Recalling that the position of equilibrium favors the weaker acid and base, the color of the solution reveals the relative strengths of the two acids.

In this experiment you are asked to determine the relative acid strengths of a number of indicators by comparing the colors of the indicators in 0.1 M solutions of known $[H_3O^+]$. The acids to be used, arranged in order of decreasing strength, are H_3O^+ (hydronium ion), CH_3CO_2H (acetic acid),

$H_2PO_4^-$ (dihydrogen phosphate), NH_4^+ (ammonium ion), and OH^- (hydroxide ion). The compositions of these known solutions and their pH values (pH= $-\log[H_3O^+]$) are summarized in Table 12.3.

TABLE 12.3 Acid and Base Strengths

Acid/Base	Composition	pH
H_3O^+	0.1 M/mole HCl/1000 mL aqueous solution.	1
CH_3COOH	Dissolve 0.1 mole of sodium acetate CH_3COONa in enough 0.1 M acetic acid (CH_3COOH) to give 1000 mL of solution.	5
$H_2PO_4^-$	Dissolve 0.1 mole NaH_2PO_4 and 0.1 mole Na_2HPO_4 in 900 mL of water. Add water to volume of 1000 mL.	7
NH_4^+	Dissolve 0.1 mole NH_4Cl in enough 0.1 M NH_4OH to give 1000 mL of solution.	9
OH^-	0.1 mole NaOH/1000 mL aqueous solution.	13

This means that, for a given number of moles of acid or base dissolved, a strong acid will result in a higher $[H_3O^+]$ and a lower pH and a strong base will result in a lower $[H_3O^+]$ and a higher pH. In a strong acid or strong base solution, the acid or base is completely dissociated. Except for the strongest acids and bases, the solutions contain quantities of the undissociated acid or base and their conjugates. This undissociated acid or base, or their conjugates will react with any strong acid or strong base introduced into the solution, minimizing the change in the pH. Solutions containing significant amounts of both acid and its conjugate base are called buffer solutions.

The use of such buffer solutions thus ensures that the observed color change is the result of reaction of the test acid or its conjugate base with the indicator base or its conjugate acid. You also will be asked to determine the pH range over which an unknown indicator changes color.

PROCEDURE

Part A

1. In a well plate, drop about 10 drops of water in five wells down the plate. Then repeat using HCl in another column of five wells. Repeat with each substance listed on Table 12.4.

The indicators used in this experiment can pose various health hazards and can cause damage to eyes and skin. Wear nitrile gloves when performing this experiment.

2. Test each substance with

   a. small piece of red litmus

   b. a small piece of blue litmus

   c. a small piece of pH paper, recording both the color and the pH (use code on vial)

   d. cresol red

   e. phenolphthalein

   f. Mg

   g. sodium bicarbonate

   h. bromothymol blue

   i. phenol red

   The reaction with $HC_2H_3O_2$ is very slow and must be carefully observed. Record all observations in Table 12.4 on the Data Sheet.

3. Take a few drops of NaOH and rub it between your finger and thumb. Record your observations in Table 12.4 on the Data Sheet.

Part B

4. Measure 10 mL of 0.10 M HCl using a small graduated cylinder. Place the 10 mL of acid into a small beaker. Record the exact volume in Table 12.6 on the Data Sheet.

Even though dilute, hydrochloric acid is corrosive and causes damage to eyes and skin. If hydrochloric acid comes in contact with skin, rinse the affected area with water for 15 minutes and alert TA.

5. Add 1–2 drops of phenolphthalein and swirl to mix.

6. Put about 10 mL of NaOH of unknown molarity into another small beaker. Fill a pipette with this NaOH.

7. Add the NaOH from the pipette drop wise into the HCl beaker. ***Count the number of drops.*** Swirl the HCl beaker after every few drops to mix in the NaOH completely.

8. Continue adding drops until the HCl solution turns light pink. This is the endpoint, where the acid has been neutralized by the base. ***Count the total number of drops*** of base used.

   20 drops = 1 mL

DATA SHEET

Name: _________________________________   Partner(s): _________________________________

TA: _________________________________   Section: ____________   Date: ______________

Part A

TABLE 12.4					
Substance	**Distilled Water**	**HCl**	**$HC_2H_3O_2$**	**NaOH**	**NH_3**
Red Litmus					
Blue Litmus					
pH Paper					
Phenolphthalein					
Bromothymol Blue					
Cresol Red					
Methyl Orange					
$NaHCO_3$					
Mg					

NaOH Observations (Step A3):

Questions and Calculations

1. List the solutions and water from lowest to highest pH.

2. Which of the two acids is more acidic?

3. Which of the two bases is more basic?

4. Which acid reacts more vigorously with magnesium? Why?

5. Using the pH recorded from the pH paper, calculate the $[H^+]$ or $[OH^-]$ of each solution.

$$pH = -\log [H^+] \qquad\qquad [H^+] = 10^{-pH}$$

$$pOH = -\log [OH^-] \qquad\qquad [OH^-] = 10^{-pOH}$$

$$pH + pOH = 14$$

TABLE 12.5			
Solution	**Acid or Base**	**pH (from Table 12.4)**	**$[H^+]$ or $[OH^-]$**
DI Water			
HCl			
$HC_2H_3O_2$			
NaOH			
NH_3			

Part B

TABLE 12.6			
Solution	**Molarity**	**Drops Used**	**Volume Used (mL)**
HCl	0.10 M		
NaOH			

Questions and Calculations

7. Write balanced equations for the reactions.

    a. hydrochloric acid and magnesium

    b. hydrochloric acid and sodium bicarbonate

    c. acetic acid and magnesium

    d. hydrochloric acid and sodium hydroxide

8. Calculate the molarity of the base, NaOH, used to neutralize the acid, HCl.

$$M_a V_a = M_b V_b$$

TA Signature

Pre-Lab Assignment ________

Safety/Participation ________

Lab Write-Up ________

Ask your TA to review your work and sign your report. The TA will sign above once satisfied that the student has performed the entire procedure. The report will not be accepted or graded unless signed.

Chemical Reagent Usage Overview
1010 Laboratories

Experiment 1: Messing with Mixtures

Only store-bought consumables are used in this activity.

Experiment 2: Density Exploration

- sugar cubes
- aspartame
- copper
- aluminum
- iron
- zinc
- glycerol
- acetone
- 2-propanol
- water

Experiment 3: Calories in Snack Foods

- cheese puffs
- marshmallows
- mixed nuts

Experiment 4: Atomic Spectra, Discharge Simulation, and Flame Tests

- 1.0 M sodium chloride
- 1.0 M potassium chloride
- 1.0 M calcium chloride
- 1.0 M barium chloride
- 1.0 M strontium nitrate
- 1.0 M lithium chloride
- 1.0 M copper chloride

Experiment 5: Chemical Bonds, Molecular Models, and Shapes

No chemical reagents used in this activity.

Experiment 6: Intermolecular Forces

- ethanol
- glycerin
- hexane
- paraffin
- mineral oil
- 2-propanol
- methanol

Experiment 7: Exploring the MOLE

- aluminum
- baking soda (sodium bicarbonate)
- vinegar (4–5% acetic acid)
- 1.0 M copper(II) sulfate
- chalk (calcium carbonate)
- bubble gum
- iron

Experiment 8: Types of Reactions

- magnesium
- sulfur
- 1.0 M hydrochloric acid
- zinc
- 1.0 M copper(II) nitrate
- copper
- 1.0 M silver nitrate
- 0.5 M sodium chloride
- 0.5 M magnesium sulfate
- 0.5 M sodium carbonate
- 6.0 M sulfuric acid
- 6.0 M hydrochloric acid
- 6.0 M sodium hydroxide
- 1.0 M acetic acid
- 1.0 M sodium hydroxide
- 1.0 M ammonium chloride
- 0.1 M barium chloride
- 0.1 M sodium sulfate
- 0.1 M sodium acetate
- 0.1 M ammonium hydroxide
- 0.1 M sodium chloride
- 0.1 M potassium nitrate

Experiment 9: Mass of a Reaction Product (Stoichiometry)

- 3.0 M hydrochloric acid
- 1.0 M sodium carbonate

Experiment 10: Limiting Reactants and Percent Yield

- copper(II) chloride
- aluminum

Experiment 11: Salt Water Concentration

- sodium chloride

Experiment 12: Properties of Acids and Bases

- bromthymol blue
- methyl orange
- cresol red
- phenolphthalein
- 0.1 M hydrochloric acid
- 0.1 M sodium hydroxide
- magnesium
- sodium bicarbonate

Lab Report Writing Suggestions

One of the most important things about science is communicating results of an experiment with other interested individuals. This is a skill that has broad applications even outside of traditional scientific fields. Throughout this semester, you will be performing experiments that require the collection of data and for observations to be made. Lab reports are the most common method for presenting your findings and discussing what they mean. This semester you will have the opportunity to write four lab reports, presenting your understanding of the performed experiments. For students in CHM 1020, this can be a daunting task as most individuals have not had all that much experience (if any) in scientific writing, let alone in the specific format of a lab report. Each report will be an opportunity to refine your skills and make tangible connections between the laboratory experiments and the larger organic chemistry content being discussed in class. The first couple reports may be a bit rough, but hopefully throughout the process you will be able to refine your skills and produce solid, well-constructed reports by the end of the semester.

The (grading) focus for each report will be slightly different:

- **Dyes and Dying:** general format, writing-style, and breadth of analysis
- **Synthesis of Aspirin:** clarity and fullness of discussion/conclusions, modifying experimental procedures
- **Paper Chromatography:** making connections to lecture content and expounding upon them
- **Extraction and Identification of DNA:** overall report quality

What follows are some dos and don'ts about lab report writing that may be useful as you are thinking about putting together yours. These are organized by section which should correspond to those sections you are including in each report. Your teaching assistants will also discuss with you those specific aspects that are important for each experiment. They are ultimately the ones who will spend the time reading and scoring the reports, so be mindful of their suggestions/requirements as well.

A Perspective on Important Aspects to Lab Report Writing

(Compiled from the experience of years of looking at organic students' lab reports.)

Introduction (~3 points)

a. Does not need to be long, roughly one paragraph long. Anything extensively more than is likely too long and superfluous.

b. It should talk generally about the concepts being covered in the lab experiment and the purpose of the experiment.

c. If any especially relevant lab techniques were utilized, they may also be introduced or described here.

d. This should be written in the present tense and absolutely should not start with the phrase: "The purpose of this experiment is …" or any variation of this.

e. Students often are tempted to lift whole passages from the lab manual. Don't do this. Points will be deducted and academic integrity violations could be instigated if especially grievous.

f. A simple reaction scheme often is included here, if appropriate (like in 1020 Experiment 5: Synthesis of Organic Compounds: Aspirin and Wintergreen Oil).

Experimental (~5 points)

a. This section should be written in the past tense and be roughly one paragraph in length with everything written in paragraph form … no bulleted lists.

b. It is not necessary to write in detail about every part of the experimental process. Things that would be common knowledge for the experimenter can be left out. (Example: You wouldn't describe how compounds were measured out.) This is a section you will likely want to have conversations with your TA about prior to turning in the report.

c. This should be written to take into account any changes or deviations that were incorporated into the procedure on lab day. It should not read as a direct copy of the lab book. You may want to reference the lab manual and document only the significant changes that were made … and that is fine if done properly with lab manual properly referenced.

Results and Discussion (~8 points)

a. This should be written in third-person past tense and likely is multiple paragraphs in length.

b. Students should critically discuss their data that might have been collected.

c. The ***Results and Discussion*** seeks describe some basic points … did the experiment achieve the expected results? What information was learned? This section should be geared toward answering questions such as these (though not explicitly). Some data collected may support or refute a particular finding, but it is important to discuss what each piece of data tells you about your experiment and why you came to that thought-process. Simply giving a bunch of results without explanation does not make for a very good discussion of results. Ideally, the ***Results and Discussion*** should account for about half the report.

d. Depending on the experiment, you may have a lot of data to work with, or very little. Don't worry so much about length. Clarity and cohesiveness is far more important than length.

e. In grading, the TAs are going to be looking at three different aspects:

    i. Presentation of data,

    ii. Demonstrated understanding of the experiment, and

    iii. Logical organization of discussion.

Conclusions (~4 points)

a. This should be approximately one paragraph in length and is usually written in present tense, though it can be done sometimes in the past tense.

b. In the ***Conclusions,*** the overall effectiveness of the experiment is described based on collected results that were just explained above. There were likely aspects that were more successful than others. This is the place to reflect upon the statements made in the ***Introduction*** and whether you approached your stated objectives for the experiments performed.

c. Also, it is advisable to give some ways in which to improve the laboratory if it were to be performed again, perhaps additional data that could have been collected or additional experiments you think might be helpful or interesting. These statements are especially important when things don't go so well with your experiment as it shows you've made some true reflections.

d. Overall, the best ***Conclusions*** sections will tie everything up, referring back to the fundamental principles discussed in the ***Introduction*** by using the results presented in the ***Results and Discussion.***

Citations

You will likely always need to cite information from a source like the lab manual. Guidelines are provided below and should be followed. Points will be deducted for failure to give appropriate references. You should consult with your TA as to their preferred in-text citation format.

Some General Things to Consider

Plagiarism

All reports should be written individually and in each students own words. Lab reports will be submitted electronically to have the originality checked. Penalties will be severe if a particularly egregious lack of originality is found.

Spelling and Grammar

Silly grammar and spelling mistakes really detract from your overall report. It's hard to take results seriously if there are a lot of spelling or grammatical errors. TAs will closely look at these aspects of your reports.

Double-Space Reports!

This makes the job of giving comments/corrections to your reports that much easier.

No First Person Tense in Lab Reports ...

Third person past tense should be used. "I" is just unacceptable in formal scientific writing. Generally speaking, the introduction should be in the present tense, the **Experimental** and **Results & Discussion** in the past tense, and the **Conclusions** perhaps either one depending on how you do it. (A quick way to lose lots of points is to use "I"s everywhere)

Write Objectively About the Experiment Performed

Make sure you have not added personal qualifications to your analysis. Statements like "the experiment went pretty well," and such, are not appropriate for scientific writing. Also, you should not write in a particularly negative fashion (i.e., whining), complaining about this or that in your conclusions or elsewhere in the lab report. Constructive means to improve the experiment are welcomed, but lack of resources or materials should not be used as reasons for why a particular experiment might have failed.

There should be a total of 20 points for each Lab Report, plus an additional 5 points for the Pre-Lab worksheet, giving a total of 25 points.

Citation Style

The ACS style is a standard method of citation in academic publications that originated with the American Chemical Society (ACS). The printed versions of the ACS style manual are entitled *ACS Style Guide: Effective Communication of Scientific Information*, 3rd ed. (2006), edited by Anne M. Coghill and Lorrin R. Garson, and *ACS Style Guide: A Manual for Authors and Editors* (1997). The guidelines may be found easily online. Shown below are some of the commonly needed formats:

Edited Book (like the lab manual)

Last Name, First Initial.; Last Name, First Initial. Title of Chapter/Experiment. In *Title of Book or Manual;* Last Name, First Initial, Ed; Publishing Company: City, Year; Volume; Pages.

Example of a Lab Manual Citation:

Alum from Aluminum Cans. In *Laboratory Guide for Chemistry.* Grossie, D.A. and Underwood, K., Eds.; Hayden-McNeil: Plymouth, MI, 2014; 7th edition; pp. 17–19.

Article Published in a Journal

Last Name, First Initial.; Last Name, First Initial. *Journal.* **Year,** *Volume*, Pages.

Example of a Journal Citation:

Deno, N. C.; Richey, H. G.; Liu, J. S.; Lincoln, D. N.; Turner, J. O. *J. Am. Chem. Soc.* **1965,** *87,* 4533–4538.

URL (web page)

Author, if available. Title of page as listed on the site. Address of page (date accessed).

Example of Web Page:

SDBS: IR (Liquid Film), benzene. http://riodb01.ibase.aist.go.jp/sdbs/cgi-bin/direct_frame_top.cgi (accessed Apr 2008).

Web Resources:

- *http://en.wikipedia.org/wiki/ACS_style*

- *http://pubs.acs.org/books/references.shtml*

- *http://library.williams.edu/citing/styles/acs.php*

- The full ACS guide is available through the WSU library class page or at: *http://pubs.acs. org.ezproxy.libraries.wright.edu/isbn/9780841239999*

Notes

Notes

Notes

Notes

Notes

Notes

Notes

Notes